101 Rare Plants of Wales
Published in Great Britain in 2019
by Graffeg Limited.

Text copyright © Amgueddfa Cymru 2019.
Written by Lauren Marrinan and Tim Rich.
With the editorial assistance of Helen Cleal.
Designed and produced by Graffeg Limited
copyright © 2019. Photo credits page 217.

Graffeg Limited, 24 Stradey Park Business
Centre, Mwrwg Road, Llangennech, Llanelli,
Carmarthenshire SA14 8YP Wales UK
Tel 01554 824000 www.graffeg.com

Amgueddfa Cymru is hereby identified as the author
of this work in accordance with section 77 of the
Copyrights, Designs and Patents Act 1988.

A CIP Catalogue record for this book is available
from the British Library.

All rights reserved. No part of this publication
may be reproduced, stored in a retrieval system
or transmitted, in any form or by any means,
electronic, mechanical, photocopying, recording
or otherwise, without the prior permission of
the publishers.

This publication is to be cited as:
Marrinan, L.M. & Rich, T.C.G. (2019).
101 Rare Welsh Plants. Graffeg.

Cover image: Purple Gromwell, photo by Tim Rich.

The publisher gratefully acknowledges the financial
support of this book by the Welsh Books Council.
www.gwales.com

ISBN 9781913134037

Printed in China

1 2 3 4 5 6 7 8 9

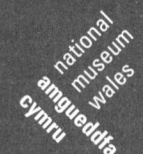

With thanks to Amgueddfa Cymru
– National Museum Wales

101 Rare Plants of Wales

Lauren Marrinan and Tim Rich

GRAFFEG

Contents

Introduction	4
Please look after our flora	6

101 Rare Welsh Plants:

1. Annual Knawel	9	28. Fen Orchid	59
2. Arctic Mouse-ear	11	29. Field Gentian	61
3. Basil Thyme	13	30. Flat Sedge	63
4. Bastard Balm	15	31. Floating Water-plantain	65
5. Beacons Hawkweed	17	32. Fly Orchid	67
6. Bog Orchid	19	Fragrant Orchids:	68
7. Bulbous Meadow-grass	21	33. Chalk Fragrant Orchid	69
8. Burnt Orchid	23	34. Marsh Fragrant Orchid	70
9. Chamomile	25	35. Northern Fragrant Orchid	71
10. Corn Buttercup	27	36. Frog Orchid	73
11. Cornflower	29	37. Glaucous Meadow-grass	75
12. Deptford Pink	31	38. Globe-flower	77
13. Divided Sedge	33	39. Goldilocks Aster	79
14. Downy Hemp-nettle	35	40. Grass-wrack Pondweed	81
15. Dune Gentian	37	41. Hairy Greenweed	83
16. Dune Wormwood	39	42. Holly-fern	85
17. Dwarf Rush	41	43. Juniper	87
18. Dwarf Spike-rush	43	44. Killarney Fern	89
19. Dyfed Hawkweed	45	45. Large-flowered Hemp-nettle	91
20. Early Gentian	47	46. Large-fruited Prickly Sedge	93
21. Early Star-of-Bethlehem	49	47. Least Whitebeam	95
Eyebrights:	50	48. Lesser Butterfly-orchid	97
22. Chalk Eyebright	51	49. Ley's Whitebeam	99
23. Cumbrian Eyebright	52	50. Llanwrtyd Hawkweed	101
24. English Eyebright	53	51. Lobed Maidenhair Spleenwort	103
25. Montane Eyebright	54	52. Marsh Clubmoss	105
26. Ostenfeld's Eyebright	55	53. Marsh Stitchwort	107
27. Welsh Eyebright	57	54. Meadow Clary	109
		55. Narrow-leaved Helleborine	111

56. Oblong Woodsia	113
57. Pale Dog-violet	115
58. Pennyroyal	117
59. Perennial Centaury	119
60. Perennial Knawel	121
61. Pillwort	123
62. Prickly Saltwort	125
63. Purple Gromwell	127
64. Purple Ramping-fumitory	129
65. Radyr Hawkweed	131
66. Red Hemp-nettle	133
67. Rock Cinquefoil	135
68. Round-leaved Whitebeam	137
69. Sea Barley	139
70. Sea Stock	141
71. Shepherd's Needle	143
72. Shore Dock	145
73. Slender Cottongrass	147
74. Slender Hare's-ear	149
75. Small-flowered Catchfly	151
76. Small-leaved Hawkweed	153
77. Small-white Orchid	155
78. Snowdon Lily	157
79. Snowdonia Hawkweed	159
80. South Stack Fleawort	161
81. Spreading Bellflower	163
82. Stag's-horn Clubmoss	165
83. Thin-leaved Whitebeam	167
84. Three-lobed Water-crowfoot	169
85. Toadflax-leaved St John's-wort	171
86. Touch-me-not Balsam	173
87. Tubular Water-dropwort	175
88. Tufted Saxifrage	177
89. Upright Clover	179
90. Upright Spurge	181
91. Viper's-grass	183
92. Welsh Groundsel	185
93. Welsh Northern Marsh-Orchid	187
94. Welsh Wood Stitchwort	189
95. White Horehound	191
96. Wild Asparagus	193
97. Wild Cotoneaster	195
98. Wood Bitter-vetch	197
99. Yellow Bird's-nest	199
100. Yellow Centaury	201
101. Yellow Whitlow-grass	203
Six outstanding plant sites in Wales	204
Organisations involved in plant conservation in Wales	208
References	210
Acknowledgements	216
Photo credits	217
Authors	219
Index	220
What you can do to help	224

Introduction

Wales has a rich and varied flora of about 1200 native and anciently-introduced flowering plants, conifers and ferns. Amongst these are many special plants ranging from arctic-alpine relicts of the last Ice Age, to recently evolved species and to medicinal herbs. They range from short-lived annual grasses to long-lived trees and vary ecologically from those growing on the sea shores to those on the tops of mountains. Some are widespread around the world and some are endemics only found in Wales.

The aim of this book is to celebrate 101 of our rarer Welsh plants and to summarise what is currently known about them. There are so many wonderful rare Welsh plants that it has been difficult to choose which to include. Our final list combined the plants given in the Welsh Government's Section 42 'Species of principal importance to Wales' (now replaced by the Environment (Wales) Act 2016 Section 7 species), with plants for which Wales has a particular responsibility (taken from Table 8 of Dines, 2008). We could probably have included another 101 with ease.

The book compliments Trevor Dine's *Vascular Plant Red Data List for Wales* (2008) – a spreadsheet-based tabulation of Welsh plants and their conservation statuses – by providing pictures and stories about the plants and what is being done to conserve them. For each species we have tried to include the following information:

- Common name, usually taken from the standard *New flora of the British Isles* by Clive Stace (2019).

- Welsh name, taken from *Enwau Cymraeg ar blanhigion* (Davies & Jones 1995).

- Latin name, also taken from Stace (2019) with the occasional addition of an out of date but widely used name.

- Conservation status, mainly taken from Dines (2008) with some updates. Note: the conservation status applies only to Wales and some species have different conservation statuses in Britain or Europe. The categories included are:

    1. **Extinct:** No reasonable doubt that the last individual has died.

2. **Extinct in the wild:** Known only to survive in cultivation, in captivity or as a naturalised population (or populations) well outside the past range.

3. **Critically Endangered:** Facing an extremely high risk of extinction in the wild.

4. **Endangered:** Facing a very high risk of extinction the wild.

5. **Vulnerable:** Facing a high risk of extinction.

6. **Near Threatened:** Close to qualifying or likely to qualify for a threatened category in the near future.

7. **Least Concern:** Relatively widespread and abundant.

8. **Data Deficient:** Inadequate information to make an assessment.

- A brief description of what the plant looks like.
- A feature of the biology, ecology or use of the plant.
- The habitat and distribution, including a map showing the distribution of records in Wales (data courtesy of the Botanical Society, Plantlife and Amgueddfa Cymru). Note that the records are plotted in 10-km squares of the national grid and there may be more than one site in each square.
- Threats and conservation.
- References for further reading. The following sources have been used for the majority of species and are not individually referenced for each account (see references for full citations): Bladwell *et al.* (2009), BSBI (2011), Clapham *et al.* (1987), Preston *et al.* (2002), Plantlife (2011), Stace (2019), Stewart *et al.* (1994) and Wigginton (1999).

We have also picked six of our favourite botanical sites in Wales where these special plants can be seen (page 204).

Please look after our flora

The wild flowers are one of Wales's most important treasures, giving enjoyment to everyone. Who cannot admire the sheets of bluebells or foxgloves in the woods and hills, or yellow fields of cowslips? But there are also many rare or easily overlooked plants which could slip away unnoticed; if nobody knows, nobody cares.

The pressures on our environment are well known – climate change and associated weather extremes, overgrazing, habitat fragmentation, soil erosion and acidification from acid rain, eutrophication from both air pollution and fertiliser drift, mineral extraction, urban development, tourism pressure, etc. – driven by technological change and our need to eat and have somewhere to live.

The flowers need somewhere to live too. Their habitats are being slowly whittled away and their populations squeezed into ever-small corners. The Welsh Rare Plants Project was an attempt to focus attention on at least some of the rare plants and provide answers, but it was only the start. At least 50 Welsh species are Critically Endangered and another 64 are Endangered (Dines 2008). Significant resources are required to deal with the scale of the problem and there are many other glamorous competing interests (red kites and red squirrels inevitably attract more attention than red hemp-nettle). The plants are equally important.

Particularly important amongst our plants are the 43 Welsh endemics which occur nowhere else in the world, 11 of which are covered here. Some of these are incredibly rare – Snowdonia hawkweed (3 plants last time it was counted) or the Beacons hawkweed (2 plants). We do not even know how many individuals of the Brecon dandelion there are, and it has only been seen recently in two sites. Ley's whitebeam is slowly declining and at the current rate will be gone by 2080. It is already too late for the extinct Robert's hawkweed, Penwyllt hawkweed and Griffith's hawkweed. Such Welsh endemics are our responsibility, nobody else's.

There are many good aspirational national policies in place to help protect the Welsh environment and there are many Sites of Special

Foxgloves in mid Wales on cleared forestry.

Scientific Interest and nature reserves. But even if the rare plants are on a nature reserve, it does not guarantee that they will survive; they need looking after too beyond simply counting how many there are (or were). For many plants, we do not even know what their precise ecological requirements are, what pollinates them or their genetic variation – all critical factors in their life cycles. Lack of appropriate management can be equally as damaging as the destructive anthropogenic factors. Retaining what we have got now is surely easier and more cost effective rather than reintroducing them once they have gone, as we have already had to do with meadow clary.

We need to do something to look after them, not just bemoan the losses. We can see the tip of the extinction iceberg looming as the Titanic of human need steams inexorably towards it; only by working together can we get the plants into the lifeboats with us.

Annual Knawel
Dinodd unflwydd

Scleranthus annuus

Welsh conservation status: Endangered

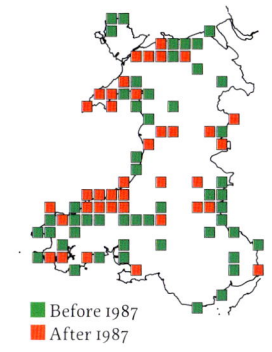

■ Before 1987
■ After 1987

A herbaceous plant which can be either annual or biennial. Flowering June–September, it has subtle green flowers which are usually self-pollinated. Large plants can produce up to 2,000 seeds.

Annual Knawel is easily overlooked as being under threat, as an initial inspection of the maps shows it is widespread throughout Wales and Great Britain. However, its significant rate of decline paints a different picture. It has been lost from perhaps half of its sites with the majority of losses since the 1950s.

Annual Knawel generally grows in two main habitats: arable fields and sandy, well-drained heathland. It has also been recorded in small soil patches within dry, open rocks.

Despite the decline, Annual Knawel is still present across most of Britain. It is rare in Ireland and it becomes commoner across its range to Central and Southern Europe. It is still scattered throughout Wales, but has been lost from many sites.

Threats and Conservation

The major factor in this species' decline is the loss of suitable arable field and heathland habitats. Current conservation action includes the requirement for its sites to be surveyed prior to the implementation of any changes and the improved management of current populations to maintain optimal conditions – the latter being particularly relevant to arable farmers.

References: Salisbury, 1961.

Arctic Mouse-ear
Clust-y-llygoden ogleddol
Cerastium nigrescens

■ Before 1987
■ After 1987

Welsh conservation status: Critically Endangered

The Arctic Mouse-ear is a small, tufted perennial which grows to just 15cm high. It is a beautiful plant, with delicate, notched white flowers that open from June to August and dark green leaves, often tinged with purple.

This plant is an arctic relic, which has persisted in cold refuges within the British Isles since the last Ice Age. Despite its common name, Arctic Mouse-ear has been recently shown not to be the same as the true Arctic species (*C. arcticum*) which does not occur as far south as Britain.

Arctic Mouse-ear typically grows in north-facing cracks and crevices of acidic mountains and cliffs, where conditions are cold and damp and vegetation is sparse, keeping competition low. It has occasionally spread to the scree or wet grasslands below the cliffs.

As it is an arctic species, this plant has a northerly distribution from Canada to Finland. It persists further south only in montane regions within Wales and Scotland. In Wales, it is confined to Snowdonia, at sites such as Clogwyn Garnedd and Clogwyn Du yr Arddu.

Threats and Conservation

Only surviving in Wales thanks to the cold climate of Snowdonia, climate change poses the biggest threat to this plant. Increases in temperature risk pushing the Arctic Mouse-ear to higher altitudes and ultimately out of its Welsh montane habitat with nowhere colder or higher to move to. Sadly, the only action that can be taken against this is to minimise human contributions to climate change, a large task that needs to be contributed to by all.

References: Brysting, 2008. Griffith, 1895.

Basil Thyme
Brenhinllys y maes

Clinopodium acinos (=Acinos arvensis)

Welsh conservation status: Vulnerable

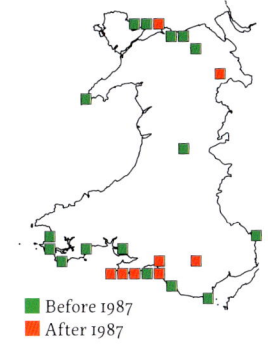

■ Before 1987
■ After 1987

A herbaceous and usually annual plant which flowers from May to September. The flowers are a beautiful violet contrasting with the small, dark green leaves. It may reach 25cm tall.

Basil Thyme was once relatively widespread and common across much of Britain. Although still widespread, this species is showing marked losses throughout its range and acts to highlight the risk facing the few remaining Welsh populations.

Basil Thyme requires open, dry habitats to thrive and is generally seen on bare grasslands or rocky ground with calcareous soil. It flourished in arable fields prior to agricultural 'improvement', but it has now been largely lost from these sites. It has also been known to occur sporadically in waste ground.

This species is widespread in the south and east of England, becoming rarer in Northern England and Wales. Basil Thyme also occurs as an introduced alien species in calcareous, gravelly sites in Ireland.

Threats and Conservation

Following its loss from arable habitats, it is now declining in grasslands due to changes in agricultural management, particularly increased weed control and the cessation of grazing. To help the recovery of this species, the intensity of the treatment on grasslands must be relaxed and traditionally successful grazing regimes reinstated.

Bastard Balm
Gwenynog

Melittis melissophyllum

Welsh conservation status: Endangered

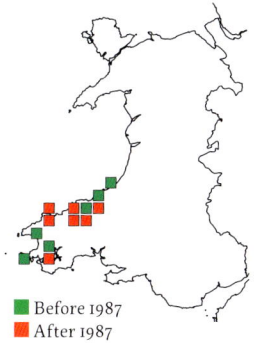

■ Before 1987
■ After 1987

The leafy stems may reach up to 70cm tall. The strongly scented flowers open May–July and the petals vary from white with pink spotting to entirely white or even purple.

At 5cm long, Bastard Balm's flowers are large for a member of the Dead Nettle family. The flowers are pollinated by long-tongued bumblebees and hawk-moths, whose tongues are able to reach the nectar at the base of the long, narrow flower tube; in the process of collecting the nectar, they are covered with pollen, which they then carry to other plants.

Bastard Balm typically occurs on woodland edges, in hedgerows and in scrub and has a preference for lightly shaded conditions.

Bastard Balm is confined to southern Britain, occurring from South-west Wales (13 sites in Pembrokeshire and one in Carmarthenshire) across England to Sussex. Globally, it is restricted to Europe and occurs south from Britain to Spain and Greece and eastwards to Russia.

Threats and Conservation

Populations of Bastard Balm in Devon and Cornwall appear stable, but decline is occurring throughout much of the rest of the British range. Intolerance of grazing is the main threat in some areas: for example, in the New Forest, plants are now found only under the cover of brambles protected from pony grazing. Habitat destruction is also playing a major role in the decline. The cultivation of material and replanting in suitable sites may prove beneficial.

References: Evans 2005. Kay & John, 1995. Pryce & Pryce 2011.

Beacons Hawkweed
Heboglys y Bannau
Hieracium breconicola

Welsh conservation status: Vulnerable

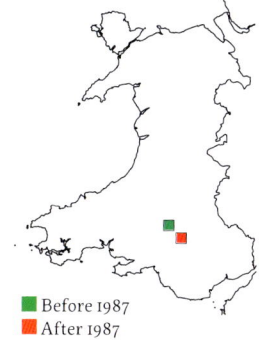

■ Before 1987
■ After 1987

This hawkweed grows up to a height of around 30cm, with the majority of leaves near the base of the stem. The vibrant yellow flowers appears from June to August.

This species has already vanished from two of its three sites, but its situation can be considered even more precarious, due to its reproductive method. Most hawkweeds do not reproduce sexually and so there is no mixing of genes between individuals. This means genetic variation becomes very low and the species is unable to adapt to new conditions, or escape from new threats upon it.

Beacons Hawkweed occurs on the outcrops of cliffs, restricted to rocks out of the reach of sheep grazing. This species is unique to Wales, having only ever been recorded at three sites, all within the Brecon Beacons, Powys. Recently it has been found to only remain at one of these, Fan Nedd, where there are only two plants.

Threats and Conservation

The decline has been attributed to sheep grazing, as hawkweeds are palatable. No conservation work is currently being undertaken to help this species and its future relies on the general protection afforded by its location within Brecon Beacons National Park. It took 8 attempts to collect seed for the Millennium Seed Bank.

References: McCosh & Rich 2018. Moore 2009.

Bog Orchid
Tegeirian bach y gors

Hammarbya paludosa

Welsh conservation status: Endangered

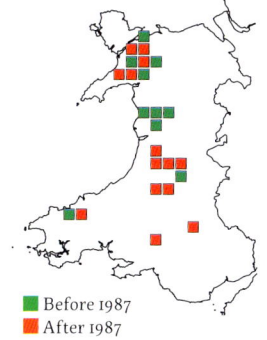

■ Before 1987
■ After 1987

Bog Orchid is a small perennial with a bulb-like base. It may be only 3cm high at its smallest, but it can be up to 12cm tall. Its flowers appear from July to September, but the small size and pale green-yellow colour means it can be difficult to spot amongst other vegetation.

Although certainly rare and declining, intensive searches recently have shown there may be more plants in Britain than once thought. It is easily over-looked due to its small size and inconspicuous appearance and its occurrence in a wet boggy habitat visited by few people.

Named after its habitat, the Bog Orchid is most commonly found in flushes, bogs and on peaty mud where the soil is permanently wet, organic and acidic. One important requirement is for movement of water through the substrate, which brings a flow of essential nutrients.

Bog Orchid occurs around the northern hemisphere, most frequently in Central and Northern Europe. Despite this, it is declining throughout Europe and is particularly rare in England and Wales.

Threats and Conservation

Bog drainage was responsible for the majority of decline prior to 1950. Since then, low or non-existent grazing levels have caused further losses by allowing the vegetation at the edge of flushes to become overgrown. In contrast, areas such as the New Forest, which uphold traditional grazing, still have good habitats and strong Bog Orchid populations. Although reintroducing grazing in areas may help the Bog Orchid, it must also be properly managed to avoid excessive levels, which lead to a risk of trampling.

Bulbous Meadow-grass
Gweunwellt Oddfog
Poa bulbosa

Welsh conservation status: Critically Endangered

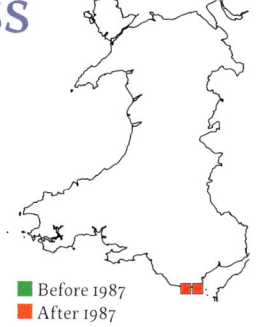

■ Before 1987
■ After 1987

A tufted grass that grows up to 40cm tall. The stems are swollen at the base and the leaves are narrow and folded. 'Flowering' occurs March–May but, instead of having flowers, it produces little 'grasslets' which eventually fall off the plant and root.

Bulbous Meadow-grass was first noted in Wales in 'considerable quantity' at The Knap, Barry in 1906. In the 1920s, construction of a promenade and gardens, and later a car park, led to removal of most of the remaining habitat. Only 40 plants remained by 2010 and only one plant in 2019. It has not been seen recently nearby at Porthkerry. Two introduced populations have been found recently in car parks.

Its typical habitat is on shingle or pebbled shores, sand dunes and sea cliffs, benefiting from soils droughted in summer which kills more competitive species. There are only a few inland occurrences.

This plant is now uncommon in Britain, restricted to England's south and east coasts and the one site in Wales. It occurs throughout much of Europe, except the most westerly and northern parts, to Asia and North Africa.

Threats and Conservation
Although the main historical threat has been from coastal development, current threats are random events affecting the last plant, visitor pressure and spread of alien plants such as Red Valerian.

Burnt Orchid
Tegeirian llosg

Neotinea ustulata (=Orchis ustulata)

Welsh conservation status: Critically Endangered

■ Before 1987
■ After 1987

This beautiful orchid grows up to about 30cm. In bud, the flowers are a deep purplish-black, but when they open they are bright white and dotted with pink-purple spots, which offers a marked contrast. This contrast gives it its name from the scorched appearance of the flowering heads.

The Burnt Orchid is a long-lived flowering plant and can take up to ten years to reach maturity and flower. Although it can reproduce by seed, this plant also creates small clusters of plants by growing new shoots from the underground tubers.

Burnt Orchid grows well in warm areas which receive good levels of sun, such as south-facing slopes. Its typical habitat is chalk or limestone grasslands which are well-grazed, with limited competition.

This species is widespread across Central Europe, north to Southern Sweden. Although scattered throughout England, it is incredibly rare in Wales, having only been seen at one site.

Threats and Conservation

In the last 50 years, there has been a decline in the numbers of Burnt Orchid, which is now restricted to a sparse scattering of sites. This has largely been due to reductions in grazing and ploughing of its habitats, as well as quarrying and construction. Wales is currently celebrating success, with the only population fruiting and setting seed in 2011, following the reintroduction of grazing and mesh caging to protect individual plants.

Reference: Foley 1990.

Chamomile
Camri

Chamaemelum nobile

Welsh conservation status: Endangered

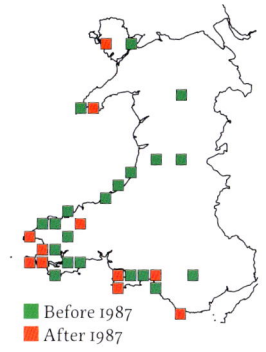

■ Before 1987
■ After 1987

Chamomile, with its characteristic scent and feathery leaves, is a perennial herb which grows stems 10-30cm long across the ground. It flowers in June–July of its second year, producing a daisy-like head of white petals, with a yellow centre.

Chamomile was historically quite widespread, thriving in the harsh conditions on heaths and common land provided by traditional grazing practices. Its medicinal uses were also well-known and it was believed to possess magical qualities, making it widely used in folklore. The demise of traditional commoning is largely responsible for the decline.

Chamomile grows in grasslands or sandy commons, coastal heaths and heathland tracks, where grazing, mowing or trampling helps keep the neighbouring plants short, the habitat open and the competition low. Its habitats are also often wet in winter.

In Europe, Chamomile populations are generally western, from Belgium to Portugal. Britain forms the northern limit and most British populations are in South Wales and England, although some introduced populations are also present in Scotland. Most Welsh populations are in Pembrokeshire and Gower, but the small populations on the Lleyn Peninsula and Anglesey are special, representing the most northerly remaining native populations in Britain.

Threats and Conservation

The loss of intense grazing and the reclamation for agriculture have caused Chamomile to be out-competed in many of its traditional habitats. Plans exist to reinstate traditional grazing methods and to link isolated populations, to encourage its spread.

Corn Buttercup
Blodyn-ymenyn yr ŷd

Ranunculus arvensis

Welsh conservation status:
Critically Endangered

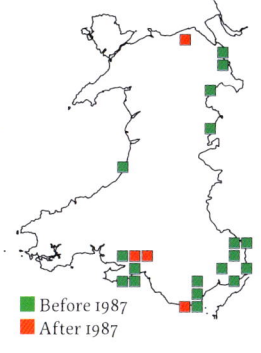

■ Before 1987
■ After 1987

The Corn Buttercup is similar in appearance to the familiar Meadow Buttercup, but is distinguishable by its smaller, paler lemon-yellow petals and its spiny seeds. It grows to 15-60cm and flowers May–June.

Corn Buttercup has been growing in arable fields since Roman times, once flourishing in low-intensity farming. Its success was due to long-lived seeds buried in the soil, which grew quickly when the ground was disturbed and conditions were at their best. Seeds were also spread with the crops.

Corn Buttercup is eliminated by herbicides and can only withstand low competition from other plants. This usually restricts it to the margins of the fields, or areas where farming intensity remains low.

It grows throughout Central and Southern Europe to North Africa and Asia. Despite once being frequent across Britain, Corn Buttercup has suffered a big decline in Wales and England and has already vanished from Ireland and the Channel Islands.

Threats and Conservation
The last 60 years of decline has resulted mainly from use of generalist herbicides and the competition caused from intensively grown crops. Low intensity farming should help this plant in the future, but options must be made available to farmers to allow them to implement this without loss of income.

Cornflower
Glas yr ŷd

Centaurea cyanus

Welsh conservation status:
Critically Endangered

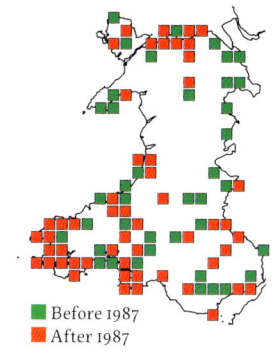

■ Before 1987
■ After 1987

Cornflower is a tall annual plant, its slender greyish stem and branches reaching up to 90cm. The eye-catching bright blue flower heads sit atop the stems and flowers open on long sunny days between June and August, to attract its bumblebee pollinators.

Ever since its arrival in Britain during the Iron Age, Cornflower was most characteristically a weed of Rye and Flax fields (rather than Corn, as the name suggests), as the seeds were adapted to live long enough to survive through less-favoured crop rotations. However, recent farming practices have pushed it from these habitats and since the 1980s it has more frequently been found in waste places, road verges and from general planting of 'wild seed' mixtures.

Cornflower is widespread throughout the temperate regions of Europe and may have originated in the South-east. Although it is scattered widely across sites in Wales and England, it rarely persists anywhere for long. Some of the records on the map may be from sites where seed has been deliberately sown.

Threats and Conservation

Cornflower has decreased due to a decline in cultivation of Rye and Flax, the introduction of seed purity regulations and the use of herbicides. It can, however, flourish once again in low intensity arable farms, which will help brighten the future of Cornflower in Britain.

Deptford Pink
Penigan y porfeydd
Dianthus armeria

Welsh conservation status: Vulnerable

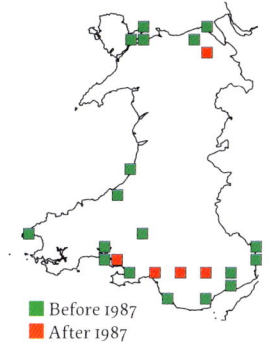

■ Before 1987
■ After 1987

Deptford Pink is a beautiful plant which can reach up to 60cm with, as the name suggests, bright pink flowers dotted with pale spots. These sit, scentless, in clusters and pairs on greyish-green stems and are open July–September.

The dramatic decline which began before 1930 has continued and Deptford Pink is now only found in 7% of its sites. The initial decline was blamed on various forms of habitat destruction, but more recently may be related to changes away from traditional grazing methods, resulting in habitats changing to unsuitable scrub and grassland.

Deptford Pink's preferred habitat is in open, dry and disturbed sites, typically margins and waysides, as well as hedgebanks and pastures. Grazing or disturbance is needed to provide suitably short vegetation, which minimises shading.

Deptford Pink occurs mainly in Western and Central Europe, but is also found in the south and east. Most British sites are in the south and about 35 remain. It has been seen in five sites in Wales since 2000.

Threats and Conservation
The decline has largely been caused by habitat loss, either through development and 'improvement', or management changes which alter conditions. On sites where the conditions have been lost, Deptford Pink could be encouraged to return by cutting back competing vegetation to reduce shade and open up the ground again, as the seeds can remain dormant in the soil for up to ten years.

Divided Sedge
Hesgen ranedig
Carex divisa

Welsh conservation status: Endangered

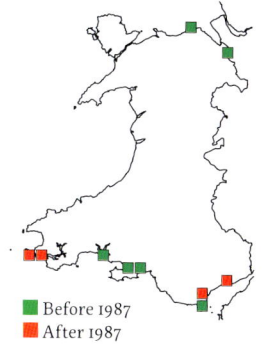

■ Before 1987
■ After 1987

A wiry-stemmed perennial plant which grows to a height of 15-80cm. The stems are topped by small, purplish-brown inflorescences in May–June, which have long leaf-like bracts growing from their base.

Divided Sedge has long roots which creep horizontally through the soil, from which new leafy stems arise. This vegetative spread of each individual produces large patches, which may result in the population looking large and healthy, but having very little genetic variation.

Divided Sedge occupies damp grasslands, dune-slacks, pasture depressions and ditches. Although able to grow in mildly salty conditions, very salty soils are not tolerated and it does not grow in salt marshes. Although it tolerates damp soils, standing water is avoided.

This is a widely distributed plant from Southern Europe to North Africa and West Asia. It becomes more and more coastal northwards and at its northern limit in Britain is nearly restricted to the coast. In Wales it occupies just a few sites along the south coast and in the north-east.

Threats and Conservation

The primary cause of decline has been habitat loss through development of coastal grasslands or their conversion to arable land. In addition, increasing controls upon tidal water incursions is reducing the salinity of nearby inland soils. The main conservation action should be to protect and maintain the conditions at the remaining sites.

Reference: Jermy et al. 2007.

Downy Hemp-nettle
Y Benboeth

Galeopsis segetum

Welsh conservation status: Extinct

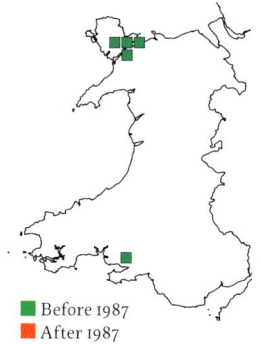

■ Before 1987
■ After 1987

An annual plant of 20-40cm height, similar in appearance to the White Dead Nettle but differing by its pale yellow flowers. These open June–August and may be tipped with purple or white. Its stem is often branched, with softly hairy leaves sprouting in opposite pairs.

Despite once being persistent within its few sites in Britain, Downy Hemp-nettle began to decline for no apparent reason between 1850s and 1890s. It was last seen in England in 1919 and finally vanished from its last Welsh site in Gwynedd in the 1970s.

Although typically found in arable cornfields within Britain, other habitats such as acid screes, gravels and scrubby woods are also home to this plant across Europe. Despite this diversity in habitat type, all habitats share the feature of having acidic, lime-deficient, freely drained soils.

Downy Hemp-Nettle grows in North-west Europe. In Britain, it was mostly characteristic of arable land in Eastern England, with a scattering of sites elsewhere, including North Wales. This distribution pattern is unusual and is unlike that of any other plant in Britain.

Threats and Conservation

Downy Hemp-nettle persisted in Britain through its characteristics as an arable weed – its seeds getting collected, resown and spread with crop seeds. However, in the 1920s, the Seeds Act forced farmers to eliminate non-crop seeds, which stopped the plant being spread in this way. Botanists have attempted resurrecting the species in Gwynedd by disturbing the ground to promote potential soil seed banks, but so far this has been unsuccessful.

References: Rich & Karran 2003.

Dune Gentian
Crwynllys Cymreig

Gentianella amarella subsp. *occidentalis*
(=*G. uliginosa*)

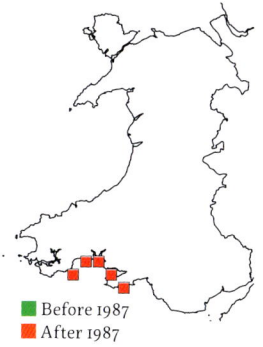

■ Before 1987
■ After 1987

Welsh conservation status: Vulnerable

A pretty annual flower, which has a rosette of leaves at the base of its reddish stem. A relatively late flowerer, the cylindrical purple flowers do not open until August–November. Plants are small, usually under 10cm and may even flower when less than 5mm tall.

The small size of Dune Gentian can cause this plant to go easily unnoticed, but its sensitivity to climate makes it even more elusive. Dune Gentian can almost disappear in wet, cool summers when soil water tables are high, only to flourish again in warmer, drier years. These dramatic fluctuations may be responsible for its recent rediscovery in North Devon and give hope for a return to its site near Tenby, where it has not been seen since 1993.

Dune Gentian is a coastal species, requiring open ground and a short sward. It is principally found within sand dunes and slacks which are nutrient-poor, dry in summer and have a diverse flora.

Dune Gentian is endemic to five dune systems in South Wales and at Braunton Burrows in North Devon, and occurs nowhere else in the world.

Threats and Conservation

The main threats to the Welsh populations are lack of grazing, which is allowing surrounding vegetation to grow and out-compete the Gentian and stabilisation of dune slack systems, which prevents new dune slacks from being formed for colonisation. Excavation of new slacks could help to create suitable habitats if coupled with the reintroduction of grazing.

References: Rich 2004. Rich et al. 2018. Rich & McVeigh 2019.

Dune Wormwood
Y feidiog ddi-sawr

Artemisia campestris subsp. *maritima*

Welsh conservation status: Critically Endangered

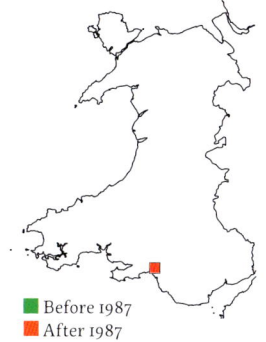

■ Before 1987
■ After 1987

This is a low-growing (to 75cm) or prostrate plant with short, finely divided, fleshy leaves on branches from the woody base. The flowers are greenish-white and are pollinated by the wind when they flower in August–September.

Debate currently surrounds this plant, concerning both its identity and its status in Britain. Subspecies maritima is a distinct coastal form of the widespread and variable Field Wormwood (*Artemisia campestris*), which differs by having more fleshy leaves with broader lobes and may be better treated as a distinct species, *Artemisia crithmifolia*. The South Wales populations are accepted as native, but the population in Crosby, England needs further investigation.

As the name suggests, this is typically a plant of sand dunes in Atlantic coastal regions. Within Britain, Dune Wormwood has been recorded since 1956 in Crymlyn Burrows, Swansea and, in 2004, plants were found at Crosby, in Merseyside. Elsewhere, Dune Wormwood occurs down the European coast from the Netherlands and south to Portugal and Spain.

Threats and Conservation

There do not appear to be any major threats to our populations, except their own rarity and low population size which make the species vulnerable to chance events: a small grass-fire just missed the Crosby population during 2007, for example. Some scrub invasion and shading from trees has affected the Crymlyn site, but is now controlled. A conservation project is underway at Crymlyn.

References: Clement 2006. Jaume et al. 2009. Smith 2008. Smith & Wilcox 2006. Twibell 2007.

Dwarf Rush
Corfrwynen

Juncus capitatus

Welsh conservation status: Endangered

■ Before 1987
■ After 1987

Dwarf Rush is a tufted annual plant, usually less than 5 cm tall. It germinates in autumn, growing through the winter and flowers March–June. It is generally green-brown in colour and turns reddish when in fruit.

The population size is variable from year to year and the plant is easily overlooked. In 1995, the population on Anglesey regarded as extinct for many years was rediscovered, with up to 20 plants. It can produce large quantities of seed for such a tiny plant.

Dwarf Rush requires a bare and open habitat, which is often achieved by grazing, but can also be regularly gained by occupying habitats which flood throughout winter months and are, conversely, droughted in summer. Examples of these include rocky outcrops, quarries, sea cliff ledges, and, in Anglesey, the margins of a sand dune hollow.

Its distribution in Europe is predominantly south-western, with scatterings throughout Central Europe and north to southern Sweden, Finland and West Russia. Dwarf Rush is unusually restricted to the west in Britain and is only present in Anglesey and on the Lizard Peninsula, Cornwall.

Threats and Conservation

Due to the requirement for open ground, decreased grazing levels have allowed more competitive plants to encroach Dwarf Rush's habitat. Some habitat has also been lost to quarrying and building development. Maintenance of remaining habitats and incorporating grazing or some other disturbance mechanisms, such as standing winter water, should help to maintain remaining populations.

Dwarf Spike-rush
Sbigfrwynen Morafon

Eleocharis parvula

Welsh conservation status: Vulnerable

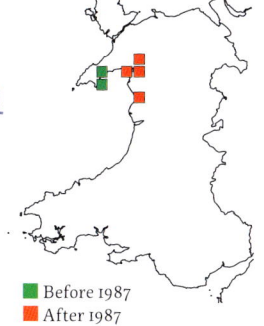

■ Before 1987
■ After 1987

An inconspicuous perennial, which only reaches up to 8cm in height, but can spread to form dense mats or patches. Both the leafless stems and flowering spikelets are greenish-brown in colour, camouflaging the plant against its muddy habitat. Flowering occurs in August–October, but is often poor and irregular.

Dwarf Spike-rush's Welsh populations are considered to be among the healthiest within Britain, helped by their long history of grazing. It mainly reproduces by small white growths, known as tubers or bulbils, budding from the tips of runners. Grazing livestock help, not only by cropping back competing vegetation, but also by spreading bulbils in mud on their feet and grazing to dislodge parts of the dense mat allowing them to float to new places with the tide.

The usual habitat is the upper tidal regions of estuaries and creeks, upon firm muddy substrate and in brackish grazing marshes. Although restricted to the upper regions to avoid strongly saline conditions, it is still typically submerged at high tide.

Dwarf Spike-rush is a European plant, scattered from Scandinavia in the north and south to Africa and Asia. In Britain it is restricted to three areas: North Wales on the coasts of Gwynedd, the Southern English coast and a couple of sites in Scotland.

Threats and Conservation
Continuity of grazing and prevention of any major disturbance should allow the remaining sites to thrive and persist. This is likely to be the situation for Welsh sites, but English sites may be more at risk. Dwarf Spike-rush has already become extinct in three European countries.

Dyfed Hawkweed
Heboglys Dyfed

Hieracium rectulum

Welsh conservation status: Data Deficient

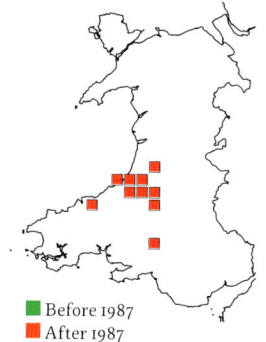

■ Before 1987
■ After 1987

This hawkweed may reach up to 50cm in height; it has a few basal leaves and up to four leaves on its stem. The flowers are bright yellow in colour and are open June–July.

Dyfed Hawkweed is currently waiting to be assigned a conservation status as, despite recent population assessments, there is uncertainty about its identity and relation to another very similar species. Variable Hawkweed, *H. submutabile*, also occurs at many of Dyfed Hawkweeds' sites and, as its vernacular name implies, is somewhat variable so it is unclear exactly how the two hawkweeds differ. Further genetic research is required before a threat category can be assigned: it is either very rare or more common.

This species is a Welsh endemic. It was originally thought to occur only on limestone rock habitats above Llangadoc in Carmarthenshire, which is where it was first discovered and described. The subsequent discovery of similar plants through much of Ceredigion revealed further possible habitats, such as acidic rock outcrops, woodland edges, roadsides, quarries and walls.

Threats and Conservation

In total about 20 populations of Dyfed Hawkweed have been reported, some of which have several thousand plants. One Carmarthenshire population, first found in 1854, was still present when visited in 1980, but has since been lost due to shading by overgrown vegetation. Although further research is required, it may be beneficial to identify this species' most vulnerable populations for potential conservation action now, rather than to wait for the taxonomy to be sorted.

References: Chater et al. 2010. McCosh & Rich 2018. Sawtschuk 2006.

Early Gentian
Crwynllys cynnar

Gentianella amarella subsp. *anglica*
(=*G. anglica*)

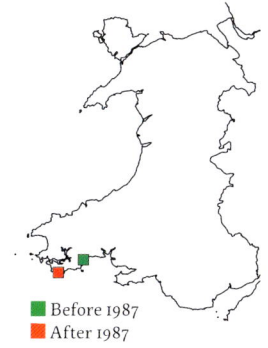

■ Before 1987
■ After 1987

Welsh conservation status: Vulnerable

The Early Gentian is a winter annual or bennial which can reach up to 20cm tall, with numerous flowers. The flowers, which open March–June, are cylindrical and a beautiful purple colour with varying tinges of white.

This is a very opportunistic species, responding well in optimal conditions. This trait results in the numbers of plants varying greatly year to year and indicates long-lived seed caches which lie dormant in the ground until conditions are right. However, what controls the germination of these seed caches is unknown.

It grows in calcareous grasslands which are well grazed with some bare ground, keeping surrounding vegetation open and no more than 5cm tall. This is often found in areas such as cliff tops, dunes and chalk downs, favouring the west- or south-facing sides of slopes, with longer periods of direct sunlight.

Early Gentian is endemic to Britain. Although once widespread in the south, it has become increasingly restricted to core chalk areas in Hampshire, Wiltshire and the Isle of Wight. Wales has just a single site on cliff-top dunes in Pembrokeshire, where it is very rare.

Threats and Conservation
Early Gentian is suffering decline throughout its range for a number of reasons, including worsening habitat quality, the loss of grazing and the effects of quarrying, ploughing and spraying. However, this is a small plant which varies in abundance and recent intensive searches have revealed new sites. Although certainly still rare, this indicates it could persist in more sites than previously thought.

References: Rich 1997. Rich et al. 2018. Rich & McVeigh 2019.

Early Star-of-Bethlehem
Seren y Creigiau
Gagea bohemica

■ Before 1987
■ After 1987

Welsh conservation status: Vulnerable

This tiny plant, often only up to 3cm, has small yellow flowers which appear (if at all) very early in the year, January–April. The plant usually has two bulbs at its base, each with two basal leaves.

Early Star-of-Bethlehem, due to its winter flowering, grew undetected in Wales until it was collected accidentally in 1965. It was initially identified from a white age-faded flower as Snowdon Lily. A second collection appeared to back this up until its hairy stem was noticed – this is not a Snowdon Lily feature. A return visit in January 1975 found fresh yellow flowers, finally allowing a correct identification.

Early Star-of-Bethlehem cannot withstand shading and needs dry conditions throughout the summer to kill competitive plants. It grows in shallow soil pockets in the cracks and ledges of rocks, or small patches of grazed grassland. By mainly growing on south- or east-facing slopes it maximises sun exposure. Flowering is usually poor in this species and in 1975, only 25 flowers were found amongst thousands of plants!

As a Southern European species, the single British site at Stanner Rocks National Nature Reserve, Powys, falls at the north-westerly limit of the range. The nearest European site is in France, isolated from the Welsh population. It is an unusual example of a drought-tolerant plant that grows well in Wales.

Threats and Conservation
Due to this plant's small size, invasive growth of larger plants and scrub need to be limited to prevent shading. It is well looked after in a National Nature Reserve and is flourishing.

Reference: Rix & Woods 1981.

Eyebrights
Effros

Eyebrights are small, branched annual plants with white to purple flowers, the petals of which have distinct upper and lower lips. The lower lip commonly has a central yellow spot. The stems and toothed leaves vary from green to red, purple or brown, depending on the species.

Eyebrights are semi-parasitic plants: their roots attach to the roots of other plant species to steal water and nutrients, but they still make their own energy using sunlight. Eyebrights are not selective about which other plant species they use as hosts. They are all characteristic of diverse, unimproved grasslands and their population sizes can vary from year to year due to sensitivity to early-season droughts.

Eyebrights are collectively referred to by the Latin name *Euphrasia officinalis* agg. which includes all 20 British Eyebright species and their 71 hybrids. The presence of hybrids between species and variations caused by different growth conditions mean it can sometimes be difficult to name individual species (Metherell & Rumsey 2018). This has led to under-recording of Eyebrights and a consequent lack of certainty in their distributions and frequencies. It remains clear, however, that Eyebrights are declining throughout their range; the six listed below are considered to be the most at threat within Wales.

Chalk Eyebright
Effros y calch

Euphrasia pseudokerneri

Welsh conservation status: Vulnerable

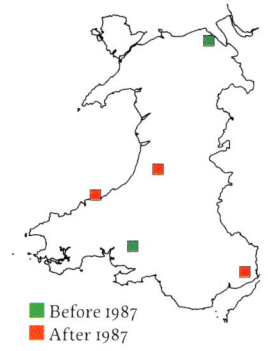

■ Before 1987
■ After 1987

Chalk Eyebright grows to a height of 20-30cm and flowers August–September to reveal a pretty white or lilac inflorescence with flowers 7-9mm long. The green leaves may also be flushed with deep red and purple.

Its main habitat is in chalk and calcareous fens, pastures and grasslands which have a diversity of other wild flowers. Chalk Eyebright is native to Britain and has not spread far, only growing elsewhere in occasional sites in France and Western Ireland. Wales is home to a few scattered sites.

Cumbrian Eyebright
Effros yr Wyddfa

Euphrasia rivularis

Welsh conservation status: Endangered

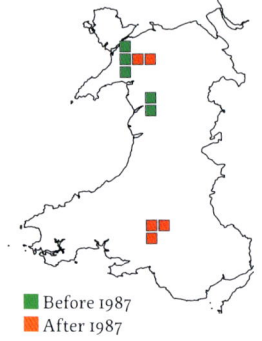

■ Before 1987
■ After 1987

The hairy and scarcely branched green or purplish stems grow to about 10cm tall, usually with lilac flowers 9mm long, which is considered large for an Eyebright. There are many glandular hairs on the stems. Flowering occurs May–July.

Cumbrian Eyebright is an upland species, living in the damp conditions of rock ledges, slopes and pastures of mountain regions, reaching its highest altitude at 750m on Snowdon. This species is unique to Britain, growing only in Carmarthenshire, Snowdonia and the Lake District.

English Eyebright
Effros Chwareog Gwalltog

Euphrasia officinalis subsp. *anglica*

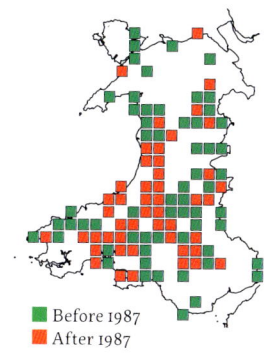

■ Before 1987
■ After 1987

Welsh conservation status: Vulnerable

This plant, although similar to the Cumbrian Eyebright, has more branches, smaller flowers and can reach 20-30cm tall. The upper lips of the flowers, which open May–September, are lilac, whereas the lower lip may be either lilac or white.

The main habitat for this species is in damp turf, heathland and pastures. It is scattered throughout the British Isles, reaching north to South-west Scotland, as well as being widespread in pastures in Wales.

Montane Eyebright
Torfagl Mynyddog

Euphrasia officinalis subsp. *monticola*

Welsh conservation status: Vulnerable

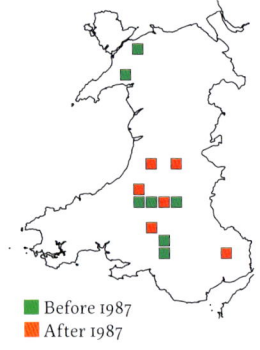

■ Before 1987
■ After 1987

The largest of the six species included here, the light green-purple stem standing at up to 35cm tall. The inflorescence is generally white, although the upper lip of the flower may be lilac. The flowers are very large (9-12.5mm long) and flowering occurs June–September.

Primarily in the montane and upland regions, Montane Eyebright grows in grasslands, hay meadows and fens. It is locally distributed, being rare in most localities, including in Gwynedd and Powys in Wales. It spreads eastwards to Russia and south to the Balkans and Carpathians.

Ostenfeld's Eyebright
Effros Ostenfeld

Euphrasia ostenfeldii

Welsh conservation status: Endangered

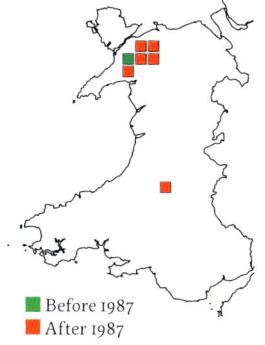

■ Before 1987
■ After 1987

This Eyebright only reaches up to 12cm in height and its leaves show a strong purple colouration on the upper surface, if not both sides, although they often look greyish, due to the dense hairs. The inflorescence is white, although the upper lip of the flowers may be lilac and the calyx-tube may also be purple-tinged or spotted. The flowering season is July–September.

Ostenfeld's Eyebright occurs in the mountains or coastal regions, on open and dry ground such as rock-ledges, cliffs, screes and sandy or stony turf. Within Britain, it is found mainly in the north-west, including North and mid Wales, but it is only found elsewhere in Iceland and the Faeroes.

Welsh Eyebright
Coreffros Cymreig

Euphrasia cambrica

Welsh conservation status: Vulnerable

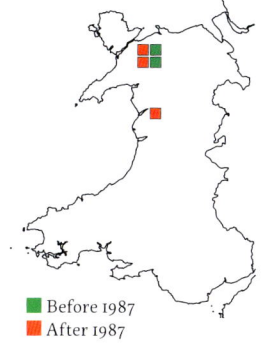

■ Before 1987
■ After 1987

The smallest of the six rare Welsh Eyebrights, reaching only 8cm in height. Its leaves are light green in colour and may be tinged with purple. The flowers, which open in July, have a white lower lip which may be yellowed and white or lilac upper lips.

This Eyebright's habitat is upon the sloping mountain grasslands of Gwynedd, where the ground is well-drained and grazed short by sheep. It may reach altitudes up to 880m. It is endemic to Wales.

Threats and Conservation
Due to the difficulties of identification and taxonomy of the Eyebrights, the threats and decline are hard to determine with certainty for individual species. Overall, Eyebrights have declined, primarily due to general habitat loss through agricultural 'improvement' and conversion of the grasslands and meadows to arable lands or uniform pastures. As with many agricultural-based threats, lower intensity management and the maintenance of remaining habitat may help to reverse the decline.

In addition, Cumbrian and Welsh Eyebrights are being threatened by the increasing temperatures of climate change, which are acting to push them out of their montane habitat, with no areas of higher altitude to take refuge in. Some management regimes have also caused decreases in Chalk Eyebright, where disturbance has promoted the growth of both a relative and competitor, the Common Eyebright (*E. nemorosa*).

References: Rich 2004. Silverside 1991.

Fen Orchid
Gefell-lys y fignen

Liparis loeselii

Welsh conservation status: Critically Endangered

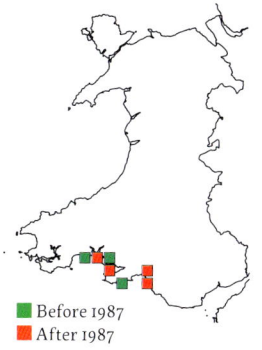

■ Before 1987
■ After 1987

This perennial orchid's stem, which can reach 20cm, has two leaves at its base and a spike of small flowers at its top. The flowers are pale yellowish-green and open June–July.

Fen Orchid has two distinct varieties which grow in different habitats. Welsh plants are known scientifically as var. (variety) *ovata* and have short, broad leaves and grow in damp, open sand in the young slacks in sand dunes. The East Anglian plants, var. *loeselii*, are narrow-leaved and grow in fens (hence the plant's common name). Despite such contrasting habitats, the conditions are actually quite similar, both being moist, poor in nutrients and diverse with other plants.

Fen Orchid is distributed widely throughout Europe, Asia and North America. Within Britain it is restricted to the South Wales coast and a few sites in East Anglia. It previously occurred in Devon.

Threats and Conservation

The East Anglian fenland populations are not faring well and have declined from 30 to 3 sites largely due to the decline of peat cutting and mowing, which has allowed scrub vegetation to encroach on the habitat.

The decline in Wales was similarly dramatic: once in eight sites with 21000 plants, it has only been seen in one site in the last ten years. Dune systems are naturally dynamic, but under-grazing and lack of disturbance have resulted in stabilised systems, preventing the creation of much-needed new slacks for Fen Orchids. Dune slack restoration at Kenfig has been successful with an increase from 400 plants in 2011 to over a thousand in 2018.

Field Gentian
Crwynllys y maes

Gentianella campestris

Welsh conservation status: Endangered

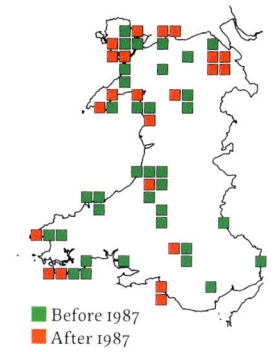

■ Before 1987
■ After 1987

A small, generally biennial herbaceous plant. The beautiful flowers, normally purple, or occasionally white or pale blue, appear July–October. This species can be distinguished from relatives by having two large and two small calyx lobes surrounding the petals.

The Field Gentian was only recently added to the UK's Biodiversity Action Plan in 2007, when the scale of its decline became apparent. It has declined dramatically over the last 100 years in the lowlands, mostly due to agricultural improvement, lack of grazing and loss of its suitable habitat. In the uplands it has been more stable and may still be locally frequent.

In lowlands, this plant grows in sand dunes, lowland pastures and on grassy heaths, whereas in the uplands it occurs on richer slopes and grasslands. Within its habitats it is able to grow on both acidic and calcareous soils.

Field Gentian has remained relatively widespread in Scotland, but most lowland British populations have been lost. In Wales it has largely retreated to coastal regions, typically sand dunes, pastures and some limestone rocks and is now rare inland.

Threats and Conservation
Although historical losses were primarily through agricultural improvement, more recently it has been attributed to reduced grazing of lowland pastures and over-grazing in uplands. In addition to reinstating traditional pasture management and decreasing upland grazing levels, plans to link good populations with new semi-natural pastures should aid its reproduction and spread.

References: Dines et al. 2005. Smith & Lockwood 2011. Rich & McVeigh 2019.

Flat Sedge
Corsfrwynen arw

Blysmus compressus

Welsh conservation status: Vulnerable

■ Before 1987
■ After 1987

Flat Sedge is a small, dark green perennial plant with wiry stems and leaves and creeping roots. Its flowers are clustered in flattened, brownish spikelets and appear from June to July.

Its typical habitats include marshes, fens, damp calcareous grasslands and margins of watercourses. These habitats are open, unshaded and usually have some have movement of base-rich ground water through the soil.

Flat Sedge grows within Europe's temperate zone and can reach as far north as Southern Scandinavia. Within Britain, it is most frequent in Northern England but occurs throughout Southern England and Scotland. In Wales, it is now confined to Powys and has not been recorded recently in Flintshire.

Despite its severe decline - it has been lost from half of its sites in which it was recorded from 1930–1960 - Flat Sedge was not made a Biodiversity Action Plan (BAP) priority species until 2007. Unless conservation action is taken soon, the decline looks set to continue.

Threats and Conservation

Due to the need for water movement within its habitats, the main threats to Flat Sedge are drainage and lowering of the water table. In addition, both over- and under-grazing pose threats, as the Flat Sedge requires a short sward to flourish well without being out-competed by neighbouring species. Current Biodiversity Action Plan measures to improve its status involve maintaining the current distribution and quality of its habitat type, as well as restoring deteriorated habitat where it remains.

Reference: Rebane 2010.

Floating Water-plantain
Dŵr-lyriad nofiadwy

Luronium natans

Welsh conservation status: Least Concern

■ Before 1987
■ After 1987

This aquatic plant has translucent, dagger-shaped submerged leaves and opaque, oval floating leaves on long stalks. The whitish petals have yellow spots at their base and float on the water surface in July–August.

This plant could be described as 'picky', only flourishing in clean waters of good quality, which allow sufficient light to reach the bottom of the water. It was once largely confined to pristine strongholds in upland Welsh lakes and the Shropshire Meres but underwent a population explosion in the canals of North-west Britain when commercial traffic ceased in the 1950s. With the resurgence of canal leisure boating in the 1970s, the water once again became muddy and it has declined dramatically.

Floating Water-plantain requires slow-flowing or relatively shallow water with good light levels and low disturbance of silts and mud. Depths of up to 2m can be tolerated but it will grow there only as rosettes of submerged leaves. Flowering is much more frequent in shallower waters, or on exposed mud when water levels drop.

Floating Water-plantain occurs only in Europe and Britain is lucky to hold the majority of its populations. Wales has a particular concentration of sites.

Threats and Conservation

Increased nutrients and sediments in waters are threatening its typical habitat, so maintaining water quality is essential. In canals, a balance must be found between its conservation needs and public leisure usage, as previously agreed limits to canal traffic have been exceeded and continue to cause decline.

Fly Orchid
Tegeirian y clêr

Ophrys insectifera

Welsh conservation status: Vulnerable

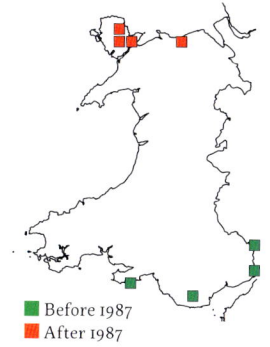

■ Before 1987
■ After 1987

With its distinctive flowers, Fly Orchid is easy to identify though spotting the slender, yellow-green plants in the first place is hard! The hanging purplish-brown flowers are scattered up the stem and rarely exceed ten in number.

The distinctive flower shape has evolved to attract digger wasps (not flies, as the common name suggests) to pollinate the plant. The flower's shape mimics the appearance of a female digger wasp and the plant produces wasp pheromones; both of these are attractive to males. As the male attempts to mate with the mimic, pollen is stuck to him and is transferred to other flowers in the same manner. Unfortunately, due to the irregularity of wasp visits, the seed set is often poor.

Fly Orchid's habitat is usually on chalk or limestones soils in partially lit, open scrub or woodland edges, but it has also been recorded in grasslands, pits, screes, railways and other habitats. The sites on Anglesey are in species-rich calcareous fens.

Although occurring across much of Central Europe, in Britain it occurs primarily in the south with scattered sites elsewhere. In Wales it is rare, with few recent records.

Threats and Conservation

Fly Orchid's decline has many contributory factors: for example, cessation of traditional coppicing has increased the density of woodland canopies, creating excessive shading. Wood clearance has also caused habitat loss for both the plant and its pollinating digger wasps and a lack of grazing of fens has increased competing vegetation. Due to the multitude of problems, a range of actions will be required to help the Fly Orchid. The continuation of traditional management techniques would help it in its remaining sites.

Reference: Karin & Karlson 1990.

Fragrant Orchids
Tegeirian pêr

Fragrant Orchids are a group of three sweetly-scented plants, which reach up to 90cm in height. Their stems have narrow leaves and the flowers are pink, with long spurs which produce nectar to attract pollinating insects. All three species grow in diverse natural grasslands, usually on calcareous soils.

Until recently, Fragrant Orchids were considered to be a single variable species, but recent genetic studies have shown there are three distinct types known as Chalk Fragrant Orchid, Marsh Fragrant Orchid and Northern Fragrant Orchid. The shape of the flowers varies between the species and they have different habitats. They are, as yet, relatively poorly studied.

Threats and Conservation

Due to the recent changes in how these three species are classified, the historical information is not specific enough and has to be interpreted with caution. This leads to difficulties in being certain how each species has fared and what has affected them, though it is clear that decline has occurred in all three for a number of reasons. Significant factors have been habitat destruction through agricultural 'improvements' such as ploughing and nutrient enrichment from fertilisers, and, in the case of Marsh Fragrant Orchid, from drainage of fens. It is hard to pinpoint where and what action needs to be taken, but the restoration of unintensive grazing in the more calcareous pastures should provide suitable conditions to halt the declines.

Reference: Campbell et al. 2007.

Chalk Fragrant Orchid
Tegeirian pêr

Gymnadenia conopsea

Welsh conservation status: Vulnerable?

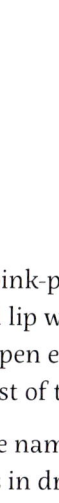

■ Before 1987
■ After 1987

The pink-purple flowers have a broad lip with three shallow lobes and open early June–July, the earliest of the three species.

As the name suggest, this species grows in dry chalk and limestone pastures. Although mainly present in Southern Britain, good populations occur as far north as County Durham and very rarely in Scotland. Although widely recorded in Wales, it is most frequent on the limestones in the north and south.

Marsh Fragrant Orchid
Tegeirian pêr y gors

Gymnadenia densiflora

Welsh conservation status: Data Deficient

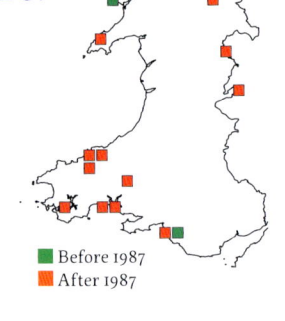

■ Before 1987
■ After 1987

Marsh Fragrant Orchid flowers are the broadest and most strongly-lobed of the three species. It is also the latest opening, the peak flowering time being July–August.

Being the 'marshy' species, this plant grows in wet meadows, fens, ditches, dune slacks and damp calcareous grasslands. Marsh Fragrant Orchid is scattered throughout Southern Britain and occurs rarely in Northern Scotland. Although uncommon in Wales, it grows within sand dunes around the coast, in the fens of Anglesey and in damp calcareous grassland inland.

Northern Fragrant Orchid
Tegeirian pêr gogleddol

Gymnadenia borealis

Welsh conservation status: Data Deficient

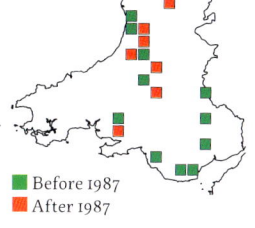

■ Before 1987
■ After 1987

This species, the smallest of the three, has flowers which have a narrow, diamond-shaped lip, which is scarcely lobed. The flowering time is mainly from late June–July.

Northern Fragrant Orchid flourishes on less strongly calcareous soils than the other two species, mainly in open grasslands and pastures and occasionally in flushes on hillsides. As indicated by the name, it is most frequent in Northern Britain and Scotland, as well as being a characteristic species of poor hill pastures in Wales.

Frog Orchid
Tegeirian y broga

Coeloglossum viride

Welsh conservation status: Endangered

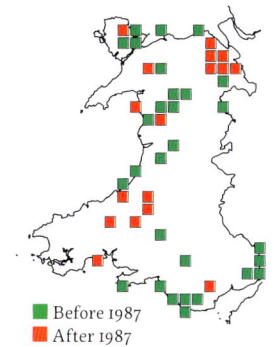

■ Before 1987
■ After 1987

The Frog Orchid gets it name from its camouflage colouration and odd flower shape, which supposedly resembles a frog! The stem, which can reach up to 40cm, is a waxy green tinged with occasional red. The delicately scented, greenish-brown flowers sprout around the stem in all directions.

In contrast to many orchids, Frog Orchid is a short-lived plant (usually only 2-5 years), with few flowers, low fruit set and a short growing season. To compensate for this, it produces seeds in large quantities. In many respects, its life-cycle is almost comparable to that of a weed!

Frog Orchid typically grows in grazed grasslands and pastures on calcareous soils, but may also be found on rocky mountain ledges and sand dunes.

Although widespread in Europe, it becomes restricted to mountain regions in the south. In Britain this plant is locally distributed and has already suffered decline in many regions, particularly Southern Scotland and Central England.

Threats and Conservation

Considerable decline has occurred both historically and recently. Much of this is due to habitat loss or destruction through agriculture such as ploughing and pasture 'improvements'. Sites with remaining populations must be conserved or protected to avoid further loss. No specific plans exist to aid its recovery, but research has shown that grazing or mowing of the habitats at the right time of year to keep surrounding vegetation short helps successful seed set and prevents it being shading out.

References: Devos et al. 2006. Willems & Melser 1998.

Glaucous Meadow-grass
Gweunwellt llwydlas
Poa glauca

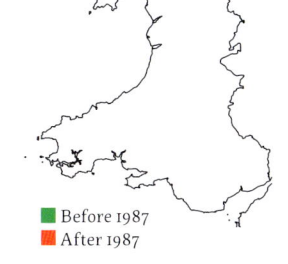

■ Before 1987
■ After 1987

Welsh conservation status: Endangered

Glaucous Meadow-grass is a tufted perennial grass, which can reach a height of 40cm. Its waxy, grey leaves are tapered and the spikelets are a deep purplish colour.

It is quite likely that Glaucous Meadow-grass is under-recorded. Its mountain habitat makes sites treacherous to survey, but its subtlety further adds to the difficulty: it typically occurs in small patches and tufts, sparsely spread and hard to see among surrounding vegetation. Its similarities to Wood Meadow-grass, a close relative which can grow in similar places, acts as an additional source of confusion.

Within its mountain habitat, Glaucous Meadow-grass is restricted to ledges, slopes and screes which are both damp and free of much other competing vegetation.

Glaucous Meadow-grass is spread across Europe to Asia and America. Within Britain, it is confined to the higher mountain regions of Snowdonia, the Lake District and Scotland.

Threats and Conservation
A decline has been observed in this species since the 1960s, although the exact reasons are uncertain and under-recording may have exaggerated the scale at which this has occurred. Glaucous Meadow-grass has been shown to be vulnerable to over-grazing, which may be the cause of some loss in the more accessible locations.

Globe-flower
Cronnell

Trollius europaeus

Welsh conservation status: Least Concern

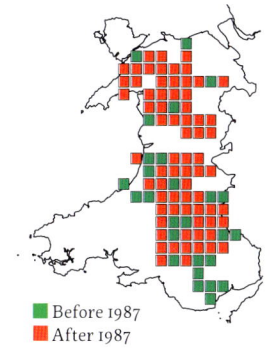

■ Before 1987
■ After 1987

Globe-flower is named after the shape of its spherical yellow flowers, which are mostly held closed and only open for up to six days in May–June. Stems can reach up to 60cm in height, each with one flower at the top.

Globe-flower is one of the few plants to have evolved a 'seed-for-seed' mutually beneficial relationship with its fly pollinators. Four closely-related fly species visit to consume pollen and nectar and then mate and lay eggs within the developing seeds. By providing this service and allowing the larvae to consume some seeds, Globe-flower ensures its own pollination as the flies go about their business visiting other Globe-flowers.

Globe-flower is a European plant, particularly widespread in the north, but more patchily distributed in the south. Within Britain, Globe-flower is also largely northern but grows throughout much of Wales.

It generally occurs in upland areas, where conditions are cool and damp. Common habitats include pastures, meadows, riverbanks, wood edges and mountain ledges.

Threats and Conservation

As agricultural management has extended into upland habitats, drainage, encroachment and fertiliser application has altered habitat conditions at the edges of Globe-flower's range and caused a reduction of its distribution. Reinstatement of traditional management should allow remaining sites to persist.

Reference: Pellmyr 1989.

Goldilocks Aster
Gold y Môr

Galatella linosyris (Aster linosyris)

Welsh conservation status: Vulnerable

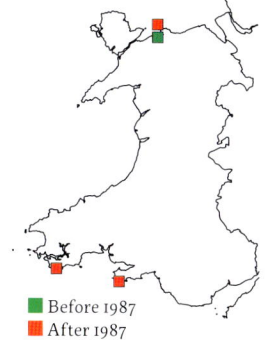

■ Before 1987
■ After 1987

A slender leafy stem, reaching up to 50cm, branches at its top to support numerous flowering heads. These flower August–September, opening with striking bright yellow florets.

Studies have shown that many Goldilocks Aster populations consist of small, isolated patches of genetically-identical plants which have arisen from one original individual. As it is self-incompatible (which means isolated clumps are unable to fertilise their own flowers), many plants in these small populations fail to produce any fertile seeds. In contrast, the large and genetically varied 7000-strong population in Pembrokeshire produces abundant seed and is thriving.

Goldilocks Aster grows in shallow soil pockets of coastal limestone cliffs and rocky slopes, sometimes spreading into cliff-top grasslands and limestone heaths. It is a poor competitor and requires short and open vegetation, usually maintained by strong coastal winds in Wales.

Goldilocks Aster is distributed primarily in South-eastern Europe and is rare in the north and west, although some small populations occur in Southern Sweden. In Britain, it is restricted to western coastal localities including Swansea, Pembrokeshire and Conwy. An old record from near Aberdaron needs confirming.

Threats and Conservation

Despite its few sites and some very small populations (such as the single remaining clump at Humphrey Head in England which may have now gone), the overall distribution of Goldilocks Aster currently seems stable. Local threats of scrub invasion can be combated using light grazing.

Grass-wrack Pondweed
Dyfrllys camleswellt

Potamogeton compressus

Welsh conservation status: Vulnerable

■ Before 1987
■ After 1987

This aquatic has a much-flattened, branched of stem up to 90cm long, which gives rise to many submerged, slender, semi-translucent leaves. It flowers and fruits sparsely in June–September and reproduction also occurs using small dense shoots, known as turions, which detach to form new plants.

The decline in Grass-wrack Pondweed has been on-going for the last 150 years and seems to be related to decreasing water quality. It appears to be one of the most sensitive aquatic plants; in Cambridgeshire its decline and extinction occurred before any obvious decline of other aquatic plants. It may thus be a useful as a sensitive indicator of water quality.

Grass-wrack Pondweed grows in lowland waterways, for example in canals, rivers, ditches and lakes. These are typically slow-flowing and rich in alkaline minerals; they also contain minerals, although not to excess.

Within Britain, the majority of records are from Central England, with just one site from Scotland. In Wales it occurs along the Montgomery Canal. Globally, it is present across Europe and eastwards to Japan.

Threats and Conservation
The decline can be largely attributed to pollution of water by excess nutrients and low light levels from mud disturbed by leisure boats. In canals, a balance must be found to allow for leisure boating with limits enforced in order to avoid unsustainable levels of usage. There are plans to help the expansion and connection of current populations and to monitor water quality to allow action where required.

Reference: Rich 2004.

Hairy Greenweed
Aurfanadl Blewog
Genista pilosa

Welsh conservation status: Vulnerable

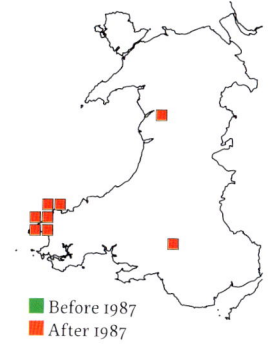

■ Before 1987
■ After 1987

A small shrub which reaches about 50cm in height, Hairy Greenweed is much branched and has numerous, small, untoothed leaves. The clustered flowers are bright yellow and open May–June.

Hairy Greenweed is unusual in the variety of sites it has colonised. These range from sea cliff heathlands in Pembrokeshire, to mountain crags on Cader Idris, limestone pavement in the Brecon Beacons and dry heaths in Southern England. It tolerates both acid and alkaline soils and grows from near sea level in Wales to high altitudes in Europe. These scattered sites with markedly different soils and climates indicate a relict distribution, now remaining from more widespread recolonisation after the last Ice Age.

Hairy Greenweed occurs scattered across Europe from Sweden to Spain. Within Britain, it is widely distributed, but is now largely coastal, having declined from inland heathlands. Vulnerability to over-grazing may restrict it to inaccessible rock ledges or ungrazed heaths.

Threats and Conservation
The loss of inland sites in the past has largely been attributed to agricultural improvements and loss of traditional management regimes causing habitat loss through scrub encroachment and/or over-grazing. The coastal populations currently appear stable, and it can occur in thousands regenerating well after fires and with light grazing.

Reference: Rich et al. 1996.

Holly-fern
Rhedynen gelyn
Polystichum lonchitis

Welsh conservation status: Vulnerable

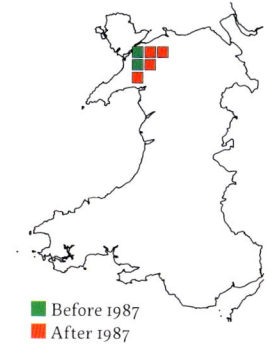

■ Before 1987
■ After 1987

Holly-fern has long, thin, dark, evergreen leaves with thin teeth along the edges of the leaf pinnules. Although the leaves appear prickly, it is a much softer and less spiny plant than its namesake, Holly.

Holly-fern was discovered in Snowdonia the 1680s by Welsh botanist Edward Llwyd. Unfortunately it subsequently fell foul of the Victorian fern collecting craze. The desire to obtain it was so great that plant hunters braved even its most treacherous mountain habitats and it is reputed that a collector died trying to collect it in 1861. Sadly, this collecting craze is largely to blame for its decline and the few plants that remain today are descended from those that survived.

Holly-fern is a montane species, preferring moist and cool conditions in base-rich crevices of rocks, ledges and stable scree, usually at altitudes above 600m.

Holly-fern grows in the cold conditions found at high altitudes in the montane regions of Europe and more widely in the arctic. The Welsh populations are the most southerly in Britain.

Threats and Conservation
With the exception to the decline caused by Victorian fern craze in the 1800s, the remaining populations seem stable and under little threat except that posed by the increasing temperatures of climate change. In Wales, the remaining populations are relatively small and sparse and should be monitored to assess any future loss.

Reference: Jones 2003.

Juniper
Merywen
Juniperus communis

Welsh conservation status: Vulnerable

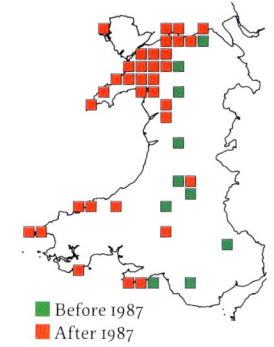

■ Before 1987
■ After 1987

Juniper is a slow-growing shrub with stiff, prickly leaves. The male and female cones open May–June and the blue-back berries ripen the following year.

Juniper was historically famous in the alcohol trade. The berries were used for flavouring gin, and, as the wood burns with invisible smoke, it was a perfect fuel for unlicensed whiskey stills.

Juniper mainly grows in grazed grasslands and open, rocky places. Other habitats include hillsides, montane regions, heaths, wind-pruned coastal cliff-tops and open woodland. It grows on both acidic and alkali soils.

Two subspecies of Juniper exist within Britain, subspecies *communis* and *nana*. Subspecies *communis* is widespread in lowland Britain and subspecies *nana* mainly in the montane areas in the north and west. Within Wales, both are uncommon and mainly occur in the north. Juniper is distributed throughout Europe, though becomes restricted to montane regions in the south.

Threats and Conservation

There has been a large decline in Juniper in the last 50 years: nearly a quarter of Welsh populations are down to a single plant and it is extinct in Carmarthenshire. Much of the decline is due to overgrazing which prevents seedlings growing. Conversely under-grazing may allow scrub and trees to shade it out. Plantlife have a conservation project aimed at maintaining the existing range and getting it to regenerate naturally.

Killarney Fern, sporophyte
Rhedynen wrychog
Trichomanes speciosum

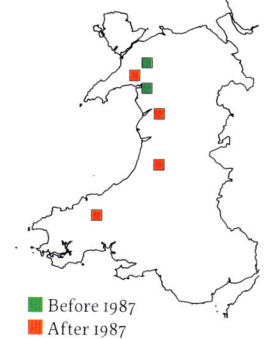

■ Before 1987
■ After 1987

Welsh conservation status: Vulnerable

A green, leafy fern which creeps horizontally to form patches with hanging fronds. The large, divided fronds are narrowly triangular in shape and are nearly translucent.

Mature plants of the Killarney Fern (called the sporophyte, as they produce spores) are very rare, as shown on the map, but the immature form (known as the gametophyte) has recently found to be relatively common and widespread. The gametophyte was completely overlooked as it looks like a mat of algae and its significance was not understood. Quite why the sporophytes are so rare but the gametophytes so widespread is currently a mystery.

Killarney Fern requires continuous heavy shade and a damp environment, commonly found in recesses, caves and cracks in woody glades and by waterfalls. These localities also provide slightly warmer refugia, protecting it from cold winter temperatures and frost, to which the Killarney Fern is very sensitive.

Within Britain, sites for the sporophyte are restricted to western coastal regions at low altitudes, where conditions are mild. Seven sporophyte localities are currently known in Wales. Killarney Fern is mainly western in Europe, but also occurs in Central Europe and the Atlantic islands.

Threats and Conservation

Sadly, it was at the hands of collecting enthusiasts that five Welsh colonies of the mature Killarney Fern were eradicated during the Victorian 'fern craze'. Another colony was lost to frosts in 1963. Killarney Fern is now under legal protection and there are plans to re-stock previously known sites.

Large-flowered Hemp-nettle
Y Benboeth amryliw
Galeopsis speciosa

■ Before 1987
■ After 1987

Welsh conservation status: Vulnerable

This beautiful weed's stem grows up to 1m high and supports several whorls of large, striking flowers, which are bright yellow with a lilac-purple lower lip. The flowers open July–September.

Large-flowered Hemp-nettle is an annual plant which is adapted to grow in the summer in areas with cool climates. The seeds are dormant when ripe and only a small proportion will germinate each year. Alternating periods of warm and cold temperature (mimicking spring) make them germinate best, but not all the seeds will germinate at once. This strategy ensures plants appear over several years, rather than all at once, giving a better chance of long term survival.

Large-flowered Hemp-nettle grows on the dark, peaty soils of arable farmland and on waste ground and disturbed roadsides. Typically it grows with root crops, for example it is commonly found in potato fields.

This species is distributed across Europe and Siberia, within the temperate and boreal regions. In Britain it is common in the north, but rare in the south. In Wales it is most frequent in the north and centre.

Threats and Conservation
The decline in Large-flowered Hemp-nettle has occurred mainly in recent decades due to modern methods of farming. Being a plant of arable land, activities such as spraying of herbicides and increased intensification are eliminating the plant and altering its habitat. Reinstatement of traditional farming methods and allowing wild flower margins on arable fields would provide suitable habitats, if these can be balanced with maintaining farming income.

Reference: Karlsson et al. 2006.

Large-fruited Prickly Sedge
Hesgen Bigog Gynnar

Carex muricata subsp. *muricata*

■ Before 1987
■ After 1987

Welsh conservation status: Critically Endangered

A densely tufted, grass-like herb, up to 25-40cm high, with green and inconspicuous flowers in May–July. The small fruits mature from greyish-green to reddish-brown.

In Wales, this species grows only near Wrexham. In 1998, felling trees as part of conservation management resulted in a population increase from two plants to over two hundred shoots in five years. However, decline has occurred again in recent years, as scrub has regrown. Historically, this site was kept open by small-scale mineral workings, but after the minerals were worked out, the disturbance stopped.

Large-fruited Prickly Sedge favours warm, dry environments, growing on steep slopes and in rocky crags and ledges, particularly on calcareous rocks. These habitats vary from open and grassy to more open woodland, providing shading remains relatively low.

This plant is present in ten scattered locations in Britain with a total population of about 800 plants. It is more widespread in Europe, with strongholds in the east and Scandinavia.

Threats and Conservation

Although shading can be tolerated at a low level, both excessive shading and excessive grazing are posing threats to this species. Cages have been successful at some sites to exclude rabbit grazing, but the resultant scrub growth adds competition, so a balance between the two must be found. Scrub clearance was conducted near Wrexham during 2011 and is being monitored to determine its success.

Least Whitebeam
Cerddin Leiaf

Sorbus minima

Welsh conservation status: Vulnerable

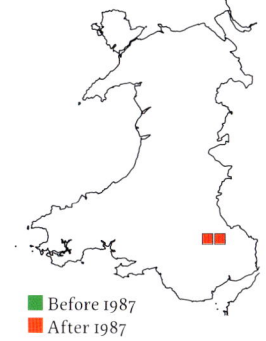

■ Before 1987
■ After 1987

At its most common height of 2-4m, Least Whitebeam is the smallest of the Whitebeam trees, although in sheltered conditions it can reach 9m. It has greyish bark and clustered white inflorescences which appear May–June. The plant fruits in August–October, producing scarlet berries, most of which contain no seeds.

The Least Whitebeam is a very special Welsh plant, growing solely in the Llangattock escarpment of the Brecon Beacons. Originally found in 1893, it is now known from just three sites and has been lost from one. Although rare in global terms, the Least Whitebeam is the most common *Sorbus* plant present at its main site, Craig-y-cilau.

This tree grows rooted into the crevices of open rocks, typically Carboniferous limestone or calcareous sandstones, where shading is minimal. Sheep grazing confines it to rocks which are out of the reach of the hungry mouths.

Threats and Conservation

The loss from Blaen Onneu is due to quarrying, as are the probable declines at other sites. At Craig-y-cilau, the population sizes in quarried areas are only around 40% of those in the un-quarried areas. After cessation of quarrying, the trees have begun to spread back onto the quarried rocks. The populations at Craig-y-cilau are well protected, as the site is a National Nature Reserve.

Reference: Rich et al. 2010.

Lesser Butterfly-orchid
Tegeirian llydanwyrdd bach

Platanthera bifolia

Welsh conservation status: Least Concern

■ Before 1987
■ After 1987

A slender and delicate plant with a light green stem, reaching up to 45cm. The numerous flowers are pale greenish-white and have a long spur at the back of the flower, which contains the nectar. The flowers are also gently scented.

Despite its common name, this species is pollinated by hawkmoths. When the hawkmoth visits the flower to drink nectar in the spur, the pollinia (sticky masses of pollen) stick to the moth's tongue and are then transferred to the next flower, enabling pollination. The tongue is one of the few places on the moth's body where there are no loose scales, allowing the pollinia to stick. Some orchids have evolved pollinia which attach to insect's eyes, where there are also no scales.

Lesser Butterfly-orchid grows in woodlands and scrub and in open meadows, grasslands and moors. It is able to cope with both acidic and alkaline soils and dry or wet conditions, such as those found in bogs and fens.

It is scattered across Southern Britain, usually in small populations, but becomes increasingly frequent towards Wales and West England. Globally, Lesser Butterfly-orchid is present throughout Europe, eastwards to Asia.

Threats and Conservation
The majority of decline occurred prior to 1930. In lowland sites, loss was attributed to habitat changes resulting from drainage, disturbance and intensification of farmland. In upland regions however, loss was mainly caused by over-grazing. The main targets to help this species are to restore low intensity management to threatened sites and to maintain existing conditions at the good populations.

Reference: Maad & Nilsson 2004.

Ley's Whitebeam
Cerddin Ley

Sorbus leyana

Welsh conservation status: Critically Endangered

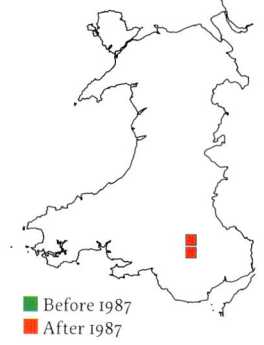

■ Before 1987
■ After 1987

Although this plant is a shrub in open conditions, when sheltered in woodland it is a small tree to at least 10m high. The bark is a dark brownish-grey and the crowded white flowers appear in May. Fruiting is sparse and sporadic and when it does occur, only a few small fruits are produced.

This species is rare, due to its recent origin as a hybrid between Rowan and Grey-leaved Whitebeam and its poor reproduction, fruiting and germination. The small fruits with few seeds limit the chances of spread or establishment and this is further reduced by deer, sheep and rabbit grazing preventing spread into neighbouring grassland and scree.

Ley's Whitebeam typically grows in scrub or open woodland on Carboniferous Limestone rock faces and cliffs. Generally it is upon the uppermost edge of the rocks, were shading is minimal.

Unique to Wales, the Ley's Whitebeam grows in two sites in the Brecon Beacons at Darren Fach and Penmoelallt. Both populations are very small, with a total of 14 natural plants in 2011, reduced to 9 (possibly 10) in 2018.

Threats and Conservation

In addition to the limitations of low seed production and sensitivity to grazing, winter frosts and grazing by caterpillars have the potential to kill or damage individual plants. Additional trees have been planted to reinforce existing populations at the Forest Nature Reserve of Penmoelallt. Darren Fach, home to eight of the shrubs, is a Brecknock Wildlife Trust Reserve. This is the flagship species of the National Botanic Garden of Wales.

Reference: Rich et al. 2010.

Llanwrtyd Hawkweed
Heboglys Llanwrtyd
Hieracium subminutidens

Welsh conservation status: Endangered

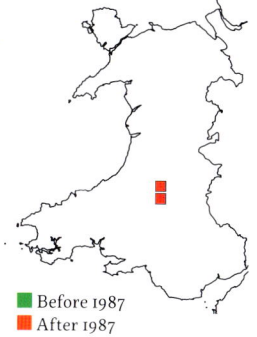

■ Before 1987
■ After 1987

This species reaches up to 70cm high, with 2-3 basal leaves and 5-7 larger leaves along the stem. The flowering season is June–July, each plant producing a branched, yellow inflorescence.

The historical records show Llanwrtyd Hawkweed occurred in six riverside sites and along a roadside bank. Surveys carried out in 2008 found 101 plants in four of the river sites. Most of the sites are very small and threatened.

Llanwrtyd Hawkweed's habitat is typically found beside rivers, growing in rocks and crevices with river gravels and sand, above the level of summer river flow. Most of the sites are partially shaded, but the plant can sometimes occur in the open.

This species is endemic to Powys, Wales. The four current sites all fall within the River Irfon catchment, near Llanwrtyd and Abergwesyn.

Threats and Conservation

Being a riverside plant, the risk of increased flooding and altered river dynamics with climate change puts the plants at risk of being washed away. Grazing pressure also poses a threat as the plants are palatable and at two sites, growth of scrub and tree results in dense shade. Localised management to limit scrub growth and grazing would be beneficial conservation efforts, but little can currently be done about the flood risk.

References: Shewring & Rich 2010. McCosh & Rich 2018.

Lobed Maidenhair Spleenwort
Duegredynen gwallt y forwyn

Asplenium trichomanes subsp. pachyrachis

■ Before 1987
■ After 1987

Welsh conservation status: Endangered

Lobed Maidenhair Spleenwort reaches up to 40cm in size, but has comparatively small and delicate leaves, with marked lobes on some of the pinnules (leaflets). It reproduces using spores, which ripen from May to October.

Despite having been known since the Victorian era, this fern's status as a subspecies has only recently been recognised. It appears naturally rare, but is puzzling botanists by its absence from many areas which would seem to suit it well.

It typically grows appressed to the surfaces of limestone rocks and walls, in places where it is sheltered and in damp or humid conditions. As a result, it is often found under overhangs, in shady outcrops and on north-facing castle walls.

Lobed Maidenhair Spleenwort is scattered across Europe, principally in the south and rarer in the northern regions. The main British stronghold lies within the Wye Valley and the largest populations grow on the walls of Chepstow and Caldicot castles.

Threats and Conservation

The biggest threat is from the 'cleaning' of castle walls to remove vegetation deemed harmful to the building's fabric; many plants were cleaned off Chepstow castle in the 1990s, though, in fact, this small fern does little damage. Both the spleenwort and the castles have a right to be conserved, highlighting the importance of finding a balance.

Marsh Clubmoss
Cnwp-fwsogl y gors
Lycopodiella inundata

Welsh conservation status: Vulnerable

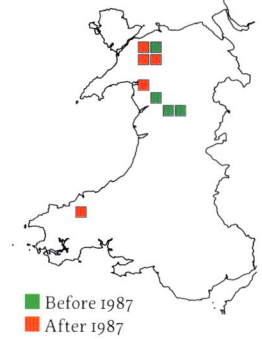

■ Before 1987
■ After 1987

A bright green, evergreen fern-relative with creeping stems. The horizontal stems branch giving rise to vertical stems up to 8cm tall, with spore-producing structures at their tips.

New plants arise from spores spread by the wind, or from vegetative division of older stems. As the stems grow slowly along the ground (2-10cm per year), they branch and root into the soil and form small patches. The branches remain green for two years, until the old stem dies and breaks into fragments, producing new separate plants.

Marsh Clubmoss requires patches of damp, open ground on bogs, heaths, lake shores and on track and path sides, where the ground is often flooded by shallow water during winter.

Marsh Clubmoss occurs mostly in Southern England and Scotland, with just a few localities between, including the sites in West and North Wales. It occurs throughout much of Europe and North America.

Threats and Conservation

Marsh Clubmoss now occupies less than a quarter of the sites it once had in Britain. In Wales, only six of the previously known 16 localities have been refound since 1987. Drainage of wetland sites has been partly to blame for losses, as has the loss of traditional grazing, resulting in the growth of taller competing vegetation.

Transplants for conservation have been successful and some new sites have been recently discovered, perhaps a result of natural spread.

Marsh Stitchwort
Serenllys llwydlas
Stellaria palustris

Welsh conservation status: Vulnerable

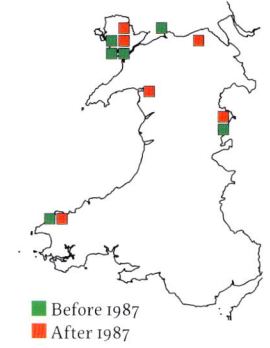

■ Before 1987
■ After 1987

A delicate and slender plant reaching around 60cm tall, with waxy-grey leaves. The five white petals are each cut nearly to their bases, allowing the green sepals beneath to be seen from above during flowering, which occurs May–July.

This species is variable in some characteristics, which can result in confusion with its British relatives, Greater Stitchwort and Lesser Stitchwort. Greater Stitchwort is most easily differentiated by having larger flowers with less cleft petals. Separating Lesser Stitchwort is trickier but it has green leaves, smaller flowers and hairy margins to the sepals.

As expected from the name, this species is found in wet and damp habitats such as fens, ditches, lake borders and marshes which shallowly flood or retain standing water throughout the winter.

This is largely a southern and eastern species within Britain: although its range does spread into Scotland and Wales, the populations there are smaller and more sparsely distributed. Marsh Stitchwort occurs across much of Northern Europe, reaching Russia in the east and southwards to the Alps.

Threats and Conservation
Although decline has been on-going for the last century, some recent studies suggest that the rate of decline may actually be small and this species is simply one which naturally grows in low abundance. The exact threats or causes of decline or rarity have not yet been established and further work is required to establish any areas which require work or conservation.

Meadow Clary
Saets y Waun

Salvia pratensis

Welsh conservation status:
Extinct in the wild

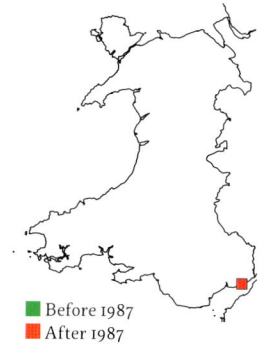

■ Before 1987
■ After 1987

A plant up to 1m in height, with a basal rosette of stalked, aromatic leaves. Each bright blue flower has a hood-like upper lip, with a projecting stigma following the curve of the hood and a three-lobed lower lip. Flowering is primarily May–July.

The name Clary originates from 'Clear-eye' and *Salvia* from Salus (Latin for health), which reflect Meadow Clary's traditional medicinal use: seeds were wetted and the resultant sticky substance (mucilage) was used to rinse troubled eyes or soothe sore throats (this is not recommended today!).

Its usual habitats are species-rich scrub edges, hay meadows and unimproved grasslands on well-drained, nutrient-poor, calcareous soils.

The only native Welsh site is in Rogiet, Monmouthshire; the western-most site of the British range. Oxfordshire is the stronghold, with 90% of the British population and it occurs in the Cotswolds and Southern England. There are other records of introduced plants in Wales and England. Although largely European, it reaches North Africa and eastwards to the Urals.

Threats and Conservation

Its decline mostly occurred across Britain prior to the 1950s, as a result of farming intensification and the loss of many hay meadows. Accidental mismanagement lead to its demise in Wales: fencing to protect it excluded grazing resulting in excessive growth of competitive plants; a subsequent grazing regime was too intense for the three remaining plants. Fortunately, seed was collected prior to extinction and seed from a single cultivated plant was crossed with plants from England for a reintroduction programme.

References: Moughan & de Vere 2012. Rich et al. 1999.

Narrow-leaved Helleborine
Caldrist gulddail
Cephalanthera longifolia

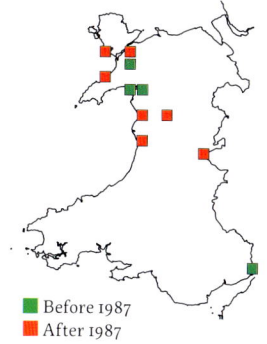

■ Before 1987
■ After 1987

Welsh conservation status: Endangered

This long-lived perennial orchid's alternative name, Sword-leaved Helleborine, aptly describes its long, blade-like leaves. Reaching a height of 40-60cm, the stem is topped with oval, white flowers with orangey-yellow tips, which open May–June.

Narrow-leaved Helleborine relies heavily upon fungi which grow associated with both its roots and the roots of trees. Two fungi groups are involved, each releasing a chemical: one stimulates germination of seed and the other stimulates its development. Large numbers of seeds are produced, but dependence on the fungi restricts where it is able to grow.

Narrow-leaved Helleborine grows in open woodland habitats or the margins of tracks, and, although preferring well-drained calcareous soils, will also grow on acid soils. Other habitats such as rocky slopes and roadsides have also been recorded.

Narrow-leaved Helleborine grows throughout Europe, to North Africa and the Far East. In Britain it is widely scattered, currently known in about 80 sites, eight of which are in Wales. Very few plants are known in half of the sites.

Threats and Conservation
Sadly some decline has been at the hands of plant collectors, the last such loss occurring in 1983. However, the majority of loss has been due to neglect of woodland management, resulting in overgrowth of vegetation and the loss of open glades. Reinstating management to open the canopy and allow light grazing should help current populations and improve sites from which loss has previously occurred.

Reference: Rumsey 2010.

Oblong Woodsia
Cor-redynen hirgul
Woodsia ilvensis

Welsh conservation status:
Critically Endangered

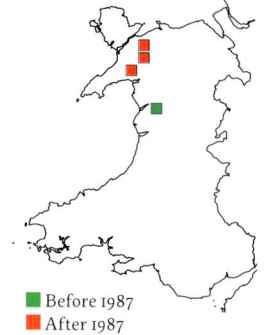

■ Before 1987
■ After 1987

An evergreen fern with tufted, hairy leaves reaching up to 15cm in length and with reddish-brown scales on their undersides. Spores are ripe July–August.

Oblong Woodsia was first found in 1690 on the rocks of Snowdonia by Welsh botanist Edward Llwyd and it always seems to have been rare. Studies have shown that its habitats have no obvious special features and it there are plenty of suitable places for it to grow in the montane areas of Britain; quite why the fern is so naturally rare is a mystery.

Its typical habitat consists of crevices and cracks of steep rock faces on mountains, where it roots into the rocks with little or no soil. These open habitats have low levels of competition and are generally freely-draining.

Within Britain, Oblong Woodsia occurs largely in Scotland although the largest colony is in the Lake District. It still grows in Gwynedd but has gone from some sites. Globally it is a relatively northern species, growing in North America and Scandinavia south to the Alps.

Threats and Conservation
Oblong Woodsia now has only eight sites within Britain, most of which are small. Much decline has been attributed to collection by fern enthusiasts during the 19th century, but, fortunately, most of the remaining plants are now within National Nature Reserves or other protected sites. Some reintroduction programmes are currently being undertaken in Scotland.

References: Aguraiuja 2011. Rich 2004.

Pale Dog-violet
Fioled welw
Viola lactea

Welsh conservation status: Vulnerable

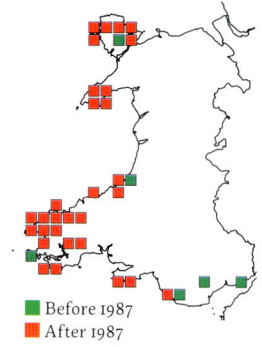

■ Before 1987
■ After 1987

A pretty plant with one or a few leafy stems growing from a basal rosette of leaves. Stems may reach 20cm in height and support creamy-violet flowers, which open May–June.

Pale Dog-violet is a poor competitor and as such requires open areas of vegetation, free from other plants, in which to establish and grow. Disturbance of the habitat – for example, by grazing or heath-fires – can open up the vegetation to create these patchy conditions, allowing seedlings to flourish.

Favouring dry and open conditions, this is a plant of lowland heath, acidic grassland and coastal cliffs. Despite the preference for dry conditions, Pale Dog-violet can also tolerate waterlogged conditions in the winter.

Pale Dog-violet is scattered mostly in South-west Britain north to Pembroke and Anglesey. A few new sites have recently been discovered in Wales. Elsewhere, it is spread down the west coast of Europe from Ireland to Portugal.

Threats and Conservation
The decline in Pale Dog-violet has mirrored the decline of its heathland habitat throughout Britain, as management regimes and land use have been changing. Many losses occurred before 1930, but more occurred after 1945 following the decline of common grazing on heathlands. Plans exist to reinstate traditional management of heathlands in order to create well-drained, patchy conditions.

Pennyroyal
Brymlys
Mentha pulegium

Welsh conservation status: Critically Endangered

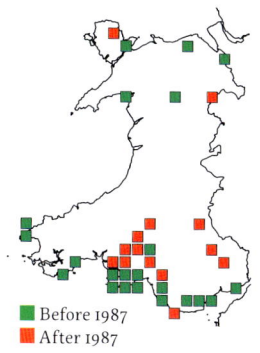

Before 1987
After 1987

An easily recognised herb with a distinct minty smell. This plant may reach 30cm in height and has whorls of lobed, lilac flowers which open August–October.

Pennyroyal has had many uses - its flowering parts can be used as an antiseptic and the pungent scent works as a flea treatment (hence the name pulegium) and a mosquito repellent. Historically, the most notorious use was to induce abortions, but this practice required near-lethal doses and resulted in many cases of toxicity and sometimes death. The availability of modern contraceptives may have meant that local or cultivated stocks around village ponds were no longer tended, leading to its decline.

Its typical habitats consist of pond edges or damp, short grasslands maintained by traditional grazing, such as those found on village greens, pastures and heaths. It occurs more rarely in woodland rides and on reservoir edges. Its habitats are usually flooded in winter.

Britain is the northern-most part of Pennyroyal's range, which extends throughout the Mediterranean to parts of Asia and Africa. Despite being once scattered in more than 200 sites across Southern Britain, the remaining populations are small and vulnerable. Some records may result from accidental introduction of seed.

Threats and Conservation
Although much decline occurred before 1950, recent changes in land management, including decreasing grazing levels, have altered the habitats conditions, leading to loss of the plant. In contrast, the New Forest continues to be traditionally managed, and, as such, has retained its Pennyroyal populations. There are plans to restore pastoral grazing regimes and to monitor all remaining sites.

References: Kay & John 1995. Khojasteh-Bakht et al. 1999. Mahboubi & Haghi 2008. Rich 2004.

Perennial Centaury
Canrhi barhaol

Centaurium portense
(=Centaurium scilloides)

Welsh conservation status: Endangered

■ Before 1987
■ After 1987

A pretty perennial plant whose flowering stems reach up to 30cm in height. The flowers occur in small clusters, with five pink petals and unusual, twisted anthers. It flowers from June–August.

Perennial Centaury occurs along the Atlantic coast, from Portugal to Wales. The British populations are somewhat distant and isolated from the European populations, so seeds may have dispersed by floating across the English Channel. Studies have shown that seeds float and remain viable for at least four weeks in sea water, so populations could have made a series of short trips to reach Cornwall, and eventually Pembrokeshire.

Perennial Centaury requires freely-draining soils on sloping coastal heaths and grassland. Open conditions are also beneficial, often provided by erosion or trampling, as well as grazing conditions.

The main British population grows near Newport, Pembrokeshire. Two populations are known in Southern Cornwall but another was lost to redevelopment. Occurrences in Kent and Sussex are probably garden escapes.

Threats and Conservation

The requirement for open conditions has meant that as traditional grazing regimes have become less common, overgrowth and scrub encroachment have pushed Perennial Centaury from some of its habitat, causing its disappearance in Cornwall. All colonies now lie within Sites of Special Scientific Interest, offering some protection, but reinstatement of appropriate coastal grazing would help boost current populations.

Reference: Rich 2005. Rich & McVeigh 2019.

Perennial Knawel
Dinodd Parhaol

Scleranthus perennis subsp. *perennis*

Welsh conservation status: Critically Endangered

■ Before 1987
■ After 1987

An inconspicuous, low-growing plant which may reach up to 20cm across. The tiny green and white flowers have five petals and open June–August.

In its only British site at Stanner Rocks National Nature Reserve, Powys, the population varies markedly in size from year to year. Only a few plants in the population flower each year and the amount of seed produced by these is also inconsistent. Consequently, the population changes dramatically between years: monitoring between 1987 and 1994 showed fluctuations between eight and 135 plants.

Perennial Knawel is a plant of rock ledges, where small pockets of soil collect. These soil pockets often dry out completely in the summer and it is highly drought tolerant.

Perennial Knawel is an unusual example of a 'continental' species in being present in Wales, presumably having colonised Wales before the countryside became covered with woodland after the last glaciation. It is mainly distributed in Central and Southern Europe to Turkey, Norway and Poland. It has declined throughout North-west Europe, but is still common in the south and east.

Threats and Conservation

Management work is being carried out at Stanner Rocks to ensure the survival of this Welsh speciality. The rocky habitats are being opened up and soils disturbed and seeds being sown to establish two new populations at the site.

Reference: Jones 2011.

Pillwort
Pelenllys
Pilularia globulifera

Welsh conservation status: Vulnerable

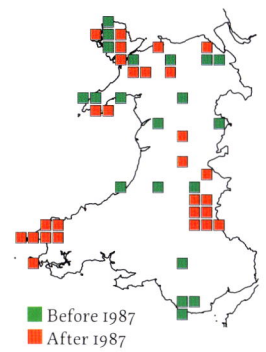

■ Before 1987
■ After 1987

A small aquatic fern, up to 8cm tall, which creeps sideways across the ground to form large patches. It is easily recognised by the coiled ends of its leaves (like a Bishop's crosier) and the brown spore-bodies ('pills', or piles!) which occur individually at the leaf bases.

Its life cycle is closely linked to its habitat - places submerged in winter and semi-exposed in summer. In summer, the drop in water levels triggers production of its spore-bodies (sporocarps). The spores are thought to be distributed to new sites on the feet of grazing animals or birds, though some can remain dormant in the bottoms of ponds and lakes.

Pillwort mainly occurs on the edges of shallow water such as around ponds and the edges of marshes and occasionally in deeper water in lakes or gravel pits. It is a highly opportunistic species, rapidly colonising bare wet substrate, which is exposed when water levels drop.

Pillwort was once more widespread but has declined throughout its range. It is typically a Western European species, although it does occur as far eastward as Finland, Poland and the Czech Republic.

Threats and Conservation
The precise cause of the decline is unknown but many sites have been lost due to habitat destruction and drainage of wetlands, as well as from the effects from nutrient-enrichment. Dredging of overgrown water bodies to create more open aquatic conditions has allowed Pillwort to recolonize old sites and to colonise new areas.

Prickly Saltwort
Helys pigog
Salsola kali subsp. *kali*

Welsh conservation status: Vulnerable

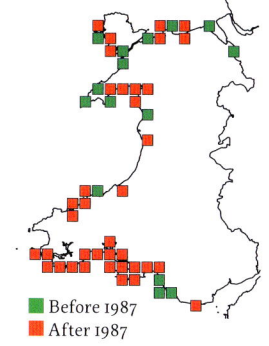

■ Before 1987
■ After 1987

A much-branched, succulent plant up to 60cm in height, with green leaves that taper spiny tips. Small inconspicuous flowers grow in the leaf axils and flower June–September.

This plant is typical of a group of annual plants that are adapted to growing quickly at the top of beaches in the summer. Seeds float and are dispersed by the tide during the winter and get washed up with other debris on the strandline, but will not germinate until after the highest spring equinox tides have finished, so they do not get washed away. Once rain has washed most of the salt from the seeds outer cases, they germinate and grow using the nutrients in the strandline debris, flowering as soon and for as long as possible in order to produce the maximum numbers of seeds. Their succulent growth form allows them to cope with salty, dry sand in the summer.

Prickly Saltwort grows on sand and shingle beaches and is present around most of the coasts of Britain and Europe.

Threats and Conservation
Although it is difficult to assess the degree of decline, due to the effects of an unpredictable, changeable habitat on population dynamics, decline has certainly been occurring since the 1960s. This is largely due to the effects of public pressure on beaches, its prickly growth form being incompatible with picnics. In addition, mechanical clearing of flotsam and jetsam and on strandlines to improve beach appearances has also removed plants, their habitats and the nutrients in the strandline.

Reference: Ignacuik & Lee 1980.

Purple Gromwell
Maenhad Meddygol

Aegonychon purpureocaeruleum
(Lithospermum purpureocaeruleum)

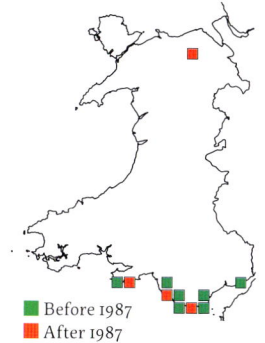

■ Before 1987
■ After 1987

Welsh conservation status: Endangered

A pretty plant, with green stems up to 60cm tall, bearing roughly-hairy leaves. Solitary purple-blue flowers grow from the leaf axils and open May–June.

In addition to propagation by seed, Purple Gromwell also spreads using creeping non-flowering stems which root where they touch the ground. When conditions are suitable, often in partial shade, it can grow very rapidly and is able to form strong populations. In Europe it is regarded as a species which likes warm, dry conditions, so it may benefit from climate change if Wales becomes warmer and drier.

It occurs in two distinct habitats in Britain, both usually on calcareous soils. Inland sites are often semi-shaded woodland edges and lane-sides. On the coast it grows amongst open scrub on slopes and rocky crags.

Purple Gromwell is a rare plant of Wales and South-west England, its original stronghold encompassed Cheddar, Bristol, Weston-Super-Mare and South Wales. In Wales, it is now known in eight native sites between Swansea and Vale of Glamorgan and one in Conwy, but has recently been lost from Monmouthshire. It occasionally escapes from cultivation elsewhere.

Threats and Conservation

The main threat to this species in its inland sites is neglect or decrease in woodland management, causing its habitats to become too dense and overgrown. Coastal sites continue to appear stable and not under any major threat. It is important for conservation of this species that inland populations are restored through reinstatement of appropriate woodland management.

Purple Ramping-fumitory
Mwg y ddaear glasgoch
Fumaria purpurea

Welsh conservation status:
Critically Endangered

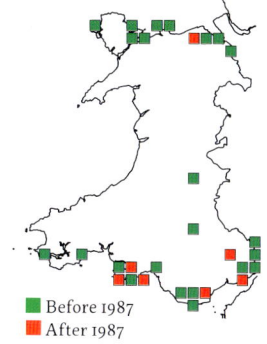

■ Before 1987
■ After 1987

This scrambling purple-pink-flowered annual may reach 1m tall and has a variable flowering season, occurring mostly March–October. The flowers are often recurved downwards and have distinctive large, oblong sepals.

Purple Ramping-fumitory is a British and Irish endemic. Although some reports indicate rarity, recent surveys suggest it may be more common, as it has long-lived seeds which may survive for up to 40 years, thus causing the plant to seemingly disappear from sites, but return erratically years later.

This species typically grows in arable fields, flower beds, hedgebanks and on walls and disturbed ground. It prefers freely-draining soils and often grows with other Fumitories. In Wales its occurrence is sporadic and it does not appear to persist anywhere for long.

Purple Ramping-fumitory is scattered across Britain, principally in the west. Although once characteristic of the Welsh Marches, it has since declined and only about ten sites are thought to remain in Wales, mainly near the coast.

Threats and Conservation
As a poor competitor against modern crops, Purple Ramping-fumitory's main threats arise from the increase in intensity of arable farming such as spraying of herbicides and conversion of arable land to pasture. Although steps can be taken to minimise this, actual evidence for decline is uncertain, due to the unpredictable population dynamics.

References: Murphy 2009. Rich 2004.

Radyr Hawkweed
Heboglys Radyr
Hieracium radyrense

Welsh conservation status: Endangered

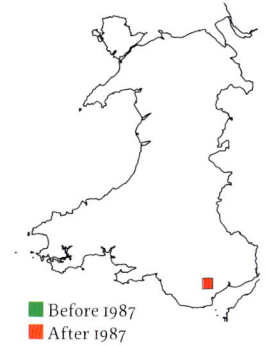

■ Before 1987
■ After 1987

A Hawkweed whose pale yellow-green stem may reach up to 80cm in height. It has a few hearted-shaped basal leaves with large teeth at their base and 2-3 stem leaves. The bright yellow flowers open May–July.

This Hawkweed was first collected in Radyr, Cardiff, over 100 years ago by H.J. Riddelsdell. It was recorded from three different sites in the Radyr area, but by 1998 was reduced to a population of about 20 plants in a garden. A new site was discovered in 2013 with 36 plants on a railway retaining wall, which was then largely destroyed in 2016 by rail maintenance work, but fortunately another 91 plants were found on the railway above. A small third new site was confirmed above Rhiwbina in 2017.

The remaining populations grows in conditions of partial shade on grassy banks, walls and on a lawn.

Threats and Conservation

With just three very small populations, this species' survival is considered highly precarious and under threat. Radyr Hawkweed is a Local Biodiversity Action Plan species in Cardiff. Material has been cultivated and seeds have been stored within a seed bank, but no legal protection currently exists. It grows well in cultivation.

References: Hutchinson & Rich 2005. McCosh & Rich 2018. Sell & Murrell 2006.

Red Hemp-nettle
Y Benboeth gulddail

Galeopsis angustifolia

Welsh conservation status:
Critically Endangered

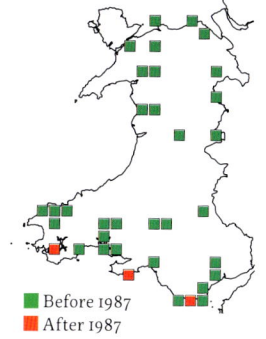

■ Before 1987
■ After 1987

A beautiful plant whose stem can reach up to 50cm and has spear-shaped leaves and distinctly-lobed, pinkish-red flowers. Flowering occurs July–October.

Traditionally, Red Hemp-nettle was a weed species of spring-sown crops and was spread as a contaminant in crop seeds. It was well-adapted for this life cycle, being able to regrow in stubble following the harvest and subsequently setting seed in autumn ready to germinate in spring. Much decline has been caused by the widespread switch to winter sown crops (to which it is not so suited), in addition to legislation controlling the occurrence of weed seeds in crop seeds.

Red Hemp-nettle grows well in warm, well-drained and calcareous soils, typically within arable field margins of spring crops. It is also present on coastal shingle and occasionally on limestone screes.

Despite once being widespread across much of England and Wales, it is now largely restricted to Southern Britain. Very few Welsh localities remain, most of which are along the south coast. Elsewhere, Red Hemp-nettle is present throughout Western, Central and Southern Europe, east to Poland.

Threats and Conservation
The intensification of farming has caused a decline in this species for over half a century, through use of competitive crops, the effects of generalist herbicides and a change to winter crop sowing. Within Wales, all arable sites have now been lost and only coastal shingle sites remain. There are plans to reintroduce seeds from these coastal sites to nearby suitable arable fields.

References: Rich & Karran 2006. Rich & Pryor 2003.

Rock Cinquefoil
Pumnalen y graig

Potentilla rupestris

Welsh conservation status:
Critically Endangered

■ Before 1987
■ After 1987

A pretty plant with a slender stem that may reach 60cm with a basal rosette of leaves and a few leaves on the stem. Five-petalled white flowers top each flowering stem, opening May–June.

Rock Cinquefoil only grows at four native sites in Britain, two of which are in Wales. It was first found at Craig Breidden, Powys in 1689, where quarrying and over-collection had a dramatic impact reducing the population to just a single plant by the 1950s. At Boughrood, Powys, it grows on shaded riverside rocks in small quantity. This poor performance in the wild is a complete contrast to its prolific growth in cultivation and it spreads rapidly in gardens; it is difficult to explain why it is quite so rare in nature.

Rock Cinquefoil grows on thin soils in rock fissures and on ledges, which are vulnerable to drought in summer months. The Boughrood site is an exception, with plants occurring on riverside rocks in an area which used to be woodland: as such, shading can cause a problem if it is not appropriately managed.

Two small sites exist in Powys and two larger and more stable populations are present in North-east Scotland. Elsewhere, Rock Cinquefoil is widespread throughout much of Western and Central Europe, ranging from Southern Sweden to the Balkans.

Threats and Conservation

Both Welsh sites are threatened. The Craig Breidden site was reduced by quarrying and numbers have now been partially boosted by transplanting individuals. At Boughrood, plants were threatened by shading from woodland, which has now been cut back. Populations should continue to improve if such conditions are maintained.

Round-leaved Whitebeam
Cerddinen Mynwy
Sorbus eminens

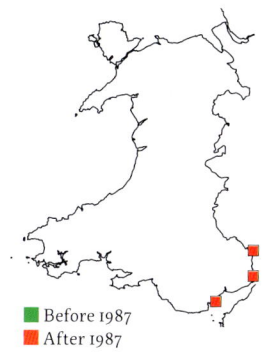

■ Before 1987
■ After 1987

Welsh conservation status: Critically Endangered

A shrub or small tree which may reach up to 15m in height. The inflorescence is white and flowering occurs May–June, followed by bright red berries in August–October. It has leaves which are often almost round, with fine regular teeth.

This species is a British endemic, growing solely England and Wales. Recent work has shown that this species is more widespread than previously thought and is now known to be scattered on limestones from the Cheddar Gorge to the Wye Valley. Only about 15 trees occur on the Welsh side of the Wye and one tree, presumably bird-sown, occurred in a nature reserve in Cardiff but has now gone. An outlying population in a park copse in Warwickshire must have been planted. The total population is thought to comprise perhaps 1000 trees.

Round-leaved Whitebeams usually occurs in open limestone habitats such as rocky scrub and grassland, cliffs, quarries and track sides, as well as in open woodland. It is one of the few whitebeams which can tolerate shade.

Threats and Conservation
The main threats are development of dense closed woodland, or uninformed woodland and scrub clearance; a full population census is required.

References: Rich et al. 2010. Rich et al. 2019.

Sea Barley
Haidd y morfa

Hordeum marinum

Welsh conservation status: Vulnerable

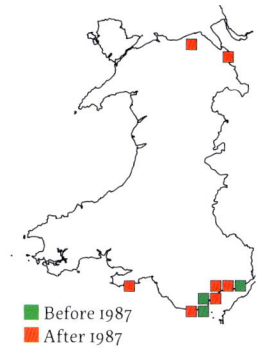

■ Before 1987
■ After 1987

A small annual grass which grows from 5–60cm tall. The dense inflorescence is composed of many spikelets which mature in colour from green to pale brown and have long awns, like cultivated barley.

Sea Barley reproduces from seeds only. At maturity the inflorescences break up and clusters of spikelets fall to the ground. One seed is present within each cluster of three spikelets. Dispersal occurs with the assistance of high tides in winter which wash the spikelets away, resulting in rows of Sea Barley plants in summer along the previous winter's strandlines.

As the name suggests, Sea Barley is largely a coastal species. It occurs on bare, brackish ground which is dry in the summer but vulnerable to winter flooding. Occasionally it may also grow in salt marshes, on waste ground or on the verges of inland roads where de-icing salt is used in winter. It is a primary coloniser and requires habitat disturbance to open up new bare areas in which to establish, as it is often out-competed by perennial grasses.

Within Wales, Sea Barley is most frequent in the south-east on the shores of the Severn Estuary. Britain is at the northern-most limit of it global range, which is from the Western European coast, throughout the Mediterranean and into Central Asia.

Threats and Conservation
Sea Barley's decline has been attributed to many factors, but is largely to the development of coastal areas, such as the rebuilding of sea defences, the filling in of coastal pools and ditches and the conversion of coastal grazing areas to arable fields. In order to conserve Sea Barley, traditional management should be reinstated, allowing natural flooding with salt water, grazing, disturbance and natural succession to resume.

Sea Stock
Murwyll arfor
Matthiola sinuata

Welsh conservation status: Vulnerable

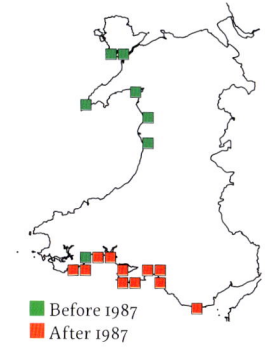

■ Before 1987
■ After 1987

A night-scented plant which can reach 1m in height with an abundance of white-hairy leaves, particularly at the base. Pretty four-petalled, pink or purple flowers a few centimetres across open June–August giving rise to long, thin fruits.

Sea Stock is relatively slow to develop, taking several years to progress from leaf-rosettes to flowering. The flowers are pollinated by nocturnal moths. When flowering is complete, the fruits split open and seeds drop from the plant and are distributed by the wind, or by the sea along the coast.

Most large populations occur on young sand dunes, where it typically grows above the drift lines or in damp un-vegetated depressions in the sand, known as blow-outs. It also occurs in small populations on sea cliffs.

Within Britain, Sea Stock is now only found on the coasts of North Devon and South Wales, but was much more widespread historically, including Ireland. Britain is at the north of the global range, which extends south along the Atlantic and Mediterranean coasts, reaching North Africa and Turkey.

Threats and Conservation
It is unclear why it has declined historically so much throughout Britain, but Sea Stock is currently undergoing a resurgence in South Wales, perhaps related to the series of warmer summers. Longer term conservation efforts may depend on establishing the causes of the historical decline.

Shepherd's Needle
Crib Gwener

Scandix pecten-veneris

Welsh conservation status: Critically Endangered

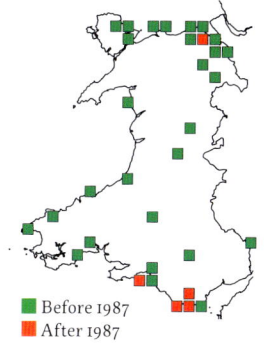

■ Before 1987
■ After 1987

A typical member of the carrot family, readily distinguished by its needle-like fruits which are present May-July. The bright green stem can reach 60cm in height and bears feathery leaves and small white flowers in umbrella-like clusters, opening April–July.

The decline of this species has been marked since the 1950s, when it was still considered an agricultural pest and was targeted using herbicides and stubble burning. The last decade has also seen some populations resurrect and there is speculation that the plant has evolved some resistance to the herbicides.

Shepherd's Needle is an arable species, occurring in the fields and margins of winter cereal crops. Occasionally plants occur on paths and banks neighbouring arable sites.

Although formerly throughout Britain, the decline has reduced it mainly to Central and Southern England and it is now uncommon in Wales. Shepherd's Needle is common and widespread around the Mediterranean.

Threats and Conservation

The methods of modern intense arable agriculture, such as spraying and seed cleaning, caused the decline of this species. Although it was once viewed as a crop pest, attitudes are changing and it is recognised as part of our wildlife heritage. Agricultural methods could be implemented that avoid both detriment to crops but also aid this species, for example, by setting aside field margins or delaying harvest until after seed production.

Reference: Rich 2004.

Shore Dock
Tafolen y traeth
Rumex rupestris

Welsh conservation status: Endangered

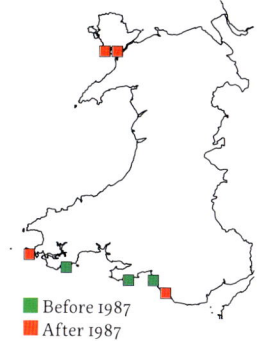

■ Before 1987
■ After 1987

Reaching up to 70cm tall, this dock has slender stems with fleshy, greyish-green leaves and dense clusters of small, greenish-red flowers. Flowering occurs June–July.

From 1985-1995, only one population of Shore Dock was known in Wales, at Newborough in Anglesey. In 1996 a population was rediscovered at Dunraven Bay, Vale of Glamorgan, which had not been seen since 1934. Three new sites have since been found in Pembrokeshire, as a result of extensive searches carried out as part of its species recovery plan. It would be nice to find more sites.

Although coastal, occurring on sand dunes, shingle beaches and sea cliffs, Shore Dock usually grows where fresh water trickles into the sea, along the cliff gullies down which fresh water is channelled, or where it accumulates in winter in dune slacks.

Shore Dock is a Western European endemic, occurring around the coasts in South-west England and western Wales and along the west coasts of France and Spain.

Threats and Conservation
In addition to the typical threats from visitor pressure and coastal development, Shore Dock also has threats linked to climate change. Increased winter storminess and raised sea levels may affect individuals growing close to the sea, particularly winter seedlings which result from the all year round germination. Competition from more hardy alien species, for example the Hottentot-fig, has also been pushing Shore Dock from its habitat. Although little can be done to combat the climatic effects, avoidance of development in known sites and maintaining fresh water flushing would be beneficial.

Slender Cottongrass
Plu'r Gweunydd Eiddil

Eriophorum gracile

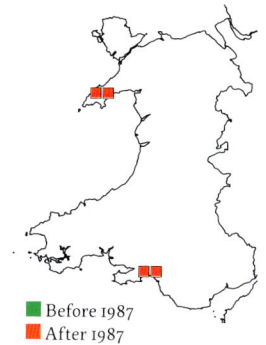

■ Before 1987
■ After 1987

Welsh conservation status: Vulnerable

Solitary stems grow up to 60cm high and are topped with clustered spikelets from May–June. During fruiting, June–July, the flower heads develop long white cottony fibres (hence its name), which help disperse seeds in the wind. It is the rarest of four cottongrasses in Wales.

This plant spreads both by seed and vegetatively, the latter of which results in quick colony formation and a tangled and extensive network of rhizomes and roots. Leafy rosettes grow from rhizomes, but die out after flowering, new rosettes arising from the ever-growing rhizome system.

Its habitat consists of the wetter parts of bogs, mires and fens, typically where there is slow movement of a water body and the resultant vegetation is a floating mat. Light shade is tolerable, but growth will not occur in strongly acidic conditions.

Always a rare, southerly species in Britain, it has declined to one site in Surrey, two in the New Forest and four in Wales, one of which is a stronghold at Crymlyn Bog. Elsewhere, Slender Cottongrass is widespread across much of Central Europe, Asia and North America.

Threats and Conservation

The decline has been attributed to drainage and infilling of wet bog habitats, which is occurring through much of the global range. However, the distribution in Wales is, at present, considered stable, with no immediate conservation plans other than to prevent drainage and development and maintain open conditions. In Surrey, open boggy habitats have been maintained using explosives!

Reference: Kay & John 1995.

Slender Hare's-ear
Paladr trwyddo eiddilddail

Bupleurum tenuissimum

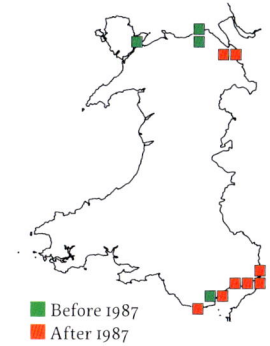

Before 1987
After 1987

Welsh conservation status: Vulnerable

This slender, yellowish-green plant may reach heights of up to 50cm. The small, sword-like leaves are up to 1cm wide. Each stem culminates in clusters of small, yellow flowers, which open July–September.

A study to see if seeds from old herbaria specimens of a range of plants could be used to resurrect extinct species found that the only ones to successfully germinate belonged to Slender Hare's-ear (2672 seeds from 26 species were tested). Each of the germinated seeds was over 100 years old and the successful germination was thought to be linked to the seeds' physiological dormancy: the embryo remains underdeveloped until germination is triggered. Such dormancy may help this annual to survive in the wild until conditions are exactly right.

Slender Hare's-ear is a plant of bare, disturbed ground where seedlings can establish in open patches. It rarely occurs inland, most sites being on coastal grasslands, banks, grazing marshes, ditch sides and sea walls.

This plant grows around the British coast from Monmouthshire to Southern Yorkshire. It occurs from Scandinavia down the coasts of Western Europe to North-west Africa and across to South-west Asia.

Threats and Conservation
Following the loss of most inland sites before 1930, the current populations and range in Britain now seem to be stable. The largest threats are the loss of grazing or disturbance, which decrease the amount of bare ground available for establishment and the loss of sea wall populations when coastal defences are developed. If those in charge of coastal management schemes are made aware of these issues, Slender Hare's-ear should continue to remain stable.

Reference: Godefroid et al. 2011.

Small-flowered Catchfly
Gludlys amryliw
Silene gallica

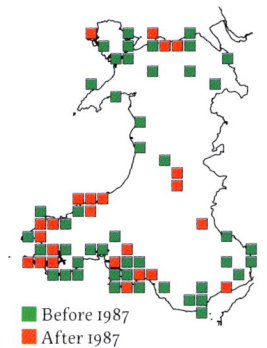

■ Before 1987
■ After 1987

Welsh conservation status: Vulnerable

A bristly annual plant up to 30cm tall. It has sticky hairs on the upper part of the plant on which insects may get stuck (hence the name 'catchfly'). Seeds germinate in the autumn, grow through the winter and spring and the flowers open June–October.

This is a genetically variable plant with a range of flower colour forms. The petals vary in colour from greenish-white and yellow to whitish-pink, pale pinks or even red and sometimes may be blotched with red. The types of hair present also varies, some lack sticky hairs, and the other hairs may spread outwards or are appressed to plant surfaces.

Usually regarded as an arable weed, its typical habitat comprises the margins of cultivated land with light sandy soils in winter- or spring-sown crops. It sometimes occurs on dry coastal banks, where soil conditions are similar to those in arable habitats.

It was once widespread in Britain, but following a 70% decline is now largely restricted to the Welsh coasts, including Pembrokeshire and Ceredigion and Southern England. It has virtually disappeared from much of Northern Europe, but is still widespread throughout Central and Southern Europe and the Mediterranean.

Threats and Conservation

Small-flowered Catchfly has declined due to use of herbicides and the competitive crop varieties used in modern intensive arable farming. In addition, coastal sites are under development pressure from tourism developments. Current conservation includes the development of arable field margin schemes and working with farmers to provide incentives for farming methods which are sensitive to the needs of the environment.

Reference: Rich 2004.

Small-leaved Hawkweed
Heboglys Mân-ddail
Hieracium angustatiforme

■ Before 1987
■ After 1987

Welsh conservation status: Critically Endangered

A small hawkweed, growing up to 40cm in height and with numerous, toothed basal leaves, but just 0-2 leaves on the stem. The yellow-rayed flowers open June–July.

Little is known about this hawkweed. It is endemic to the Brecon Beacons, Wales and has been recorded in four sites on cliff ledges, rocky screes, grass banks and streamsides. It has only been seen recently from a single site at Craig Cerrig-gleisiad National Nature Reserve.

Threats and Conservation
Over-grazing by sheep has posed a threat to most species of hawkweed as they are typically very palatable to sheep. Sheep were removed from Craig Cerrig-gleisiad and replaced by low intensity cattle-grazing with no sheep in the early 1990s; the populations of all hawkweeds there are now flourishing. Surveys are needed to search for and assess the conservation needs of this rare hawkweed.

Reference: McCosh & Rich 2018. Sell & Murrell 2006.

Small-white Orchid
Tegeirian bach gwyn

Pseudorchis albida (=Leucorchis albida)

Welsh conservation status: Critically Endangered

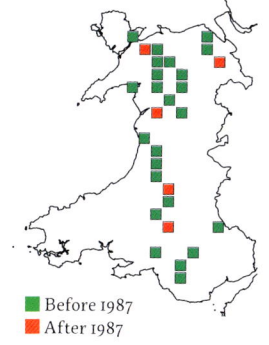

■ Before 1987
■ After 1987

A beautiful plant with a long, shiny, green stem and sheathing leaves. Small white-green flowers form a cylindrical inflorescence on the upper part of the stem. Flowering occurs May–July, but is brief for individual flowers, the lower flowers usually fade before those further up open.

Small-white Orchid is a variable species throughout its range. Some taxonomists consider that there are three subspecies: subsp. *albida*, subsp. *straminea* and subsp. *tricuspis* but other studies have shown few clear distinctions in areas where the ranges overlap. If considered as separate subspecies, subsp. *albida* is the only one present in Britain.

The habitat is not easy to classify, as Small-white Orchid regularly grows in various different vegetation types with short open swards and in the transitions between them. It has been recorded in poor hill pastures, stream sides, cliffs, heather moorland and montane grasslands.

Within Britain, this species is mostly present in Scotland, with a few scattered sites across mid and North Wales and Northern England. Its global range depends on the taxonomic view, but subsp. *albida* ranges eastwards to Scandinavia and Russia, south to montane Spain and to Eastern Europe.

Threats and Conservation

The decline in Britain is estimated to be a 65% loss of sites, largely though habitat destruction including the 80% of heathland lost due to conversion to arable, quarrying, forestry and development. Overgrazing and scrub encroachment have also contributed to the decline. In order to conserve remaining populations, management is required which allows disturbance, to maintain a short sward and prevention of further habitat loss or fragmentation.

References: Jersáková et al., 2011.

Snowdon Lily
Lili'r Wyddfa

Gagea serotina (=Lloydia serotina)

Welsh conservation status: Vulnerable

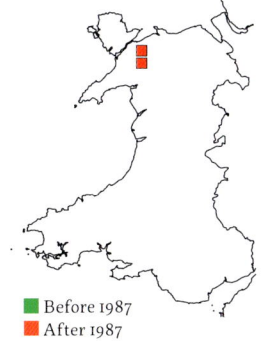

■ Before 1987
■ After 1987

A delicate alpine up to 15cm tall with grass-like leaves and one, occasionally two or rarely four flowers on one stem. The petals are white with purple veins and yellowish bases and the flowers open between May and June.

Following its discovery by Edward Llwyd late in the 17th century, Snowdon Lily has become an iconic Welsh plant – widely heard of but rarely seen. Small populations are present on six or seven cliffs in Snowdonia. Genetic studies show the Snowdonia plants have been isolated from the nearest European populations in the Alps for a long time by the Ice Ages and that some of the smallest populations have lower seed production, either due to this reduced genetic variation, or due to suboptimal ecological conditions in Wales.

In Wales, Snowdon Lily grows in damp cracks and crevices, mostly on north-facing mountain cliffs. This differs from the sunny alpine tundra slopes it inhabits in the Alps and the Rocky Mountains.

Snowdonia is the only place in Britain with this plant, but it has a wide range in the northern hemisphere, from North America to China and arctic Russia. Surprisingly, it is absent from Scandinavia.

Threats and Conservation
Historically, many plants were collected from accessible places, but protection by law means this is now an offence. Grazing continues to constrain populations to inaccessible locations. Loss to potential rock falls is also a threat. Careful management and grazing restrictions may even help to minimise loss due to climate change, which threatens to push many of Snowdonia's plants from their habitats as temperatures rise. Snowdon Lily currently faces a rather uncertain future.

References: B Jones 2001, 2010. Jones & Gliddon 1999. Jones et al. 2001.

Snowdonia Hawkweed
Heboglys Yryri
Hieracium snowdoniense

Welsh conservation status: Critically Endangered

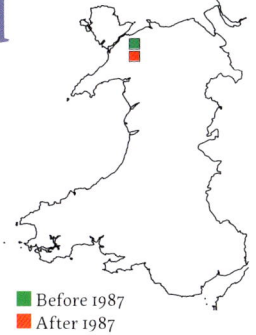

■ Before 1987
■ After 1987

The long, green stem has a rosette of leaves at its base and is topped with a branching cluster of golden-yellow flowers. Flowering is mainly in July, but many mountain hawkweeds can have irregular or unpredictable flowering patterns and this is the case for Snowdonia Hawkweed.

Historically, Snowdonia Hawkweed has been recorded in seven localities. Surveys in the 2000s have found it has declined greatly, down to only three plants in one site, making it one of the rarest plants in the world.

As its name suggests, this Hawkweed resides in Snowdonia and is a mountain plant growing on the base-rich soils on rocks and crags. Hawkweeds are very palatable to sheep, restricting the habitat to only those rocks out of the reach of hungry mouths.

All sites, both current and historical, are in Snowdonia. The last remaining site is in Cwm Idwal National Nature Reserve.

Threats and Conservation
Snowdonia Hawkweed is being cultivated at the National Botanic Garden of Wales, with a view to restocking the populations in Snowdonia. Sheep have been removed from grazing Cwm Idwal, which is having a beneficial effect on the entire flora and may allow the Snowdonia Hawkweed to spread naturally. Acid rain may also be reducing the suitability of the soils, by leaching out the much needed base-rich minerals and creating acidic conditions.

References: McCosh & Rich 2018. Rich 2003.

South Stack Fleawort
Chweinllys Arfor

Tephroseris integrifolia subsp. *maritima*

Welsh conservation status: Vulnerable

■ Before 1987
■ After 1987

A plant which typically reaches 25-60cm with large, cabbage-like hairy leaves. The daisy-like bright yellow flowers, of which there may be up to 12, open May–July, although only 40% of individuals may flower each season.

This plant has evolved as a distinct form adapted to sea cliffs in Wales. Its close relative, the Field Fleawort, is an uncommon plant of short chalk and limestone grasslands in England, but quite how or when it got to Anglesey and adapted to the very different maritime climate is unknown. It was first discovered in 1813 and was named as a subspecies of Field Fleawort in the 1860s.

It occurs on sea cliffs, ledges and rock crevices in maritime grasslands, growing best where protected from the prevailing winds. Having only been recorded in small colonies upon the cliffs of Holy Island, Anglesey, it is not only its small numbers making it so special to Wales, but also its exclusivity; it occurs nowhere else in the world.

Threats and Conservation

South Stack Fleawort is well-protected in nature reserves and the plant is regularly monitored and shows no signs of decline.

Spreading Bellflower
Clychlys ymledol
Campanula patula

Welsh conservation status: Critically Endangered

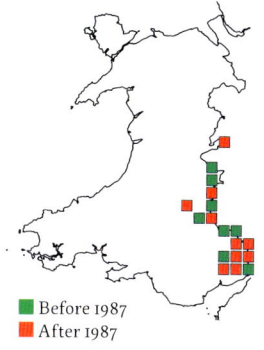

■ Before 1987
■ After 1987

A much branched plant with slender stems up to 60cm in height, bearing bright blue, star-shaped flowers which open July–November. As the name suggests, the petals are splayed out widely compared to other Bellflowers.

Studies at the National Botanic Garden of Wales show that as it has declined since the early 1800s, the geographic range has contracted and fragmented and there has been an overall loss of genetic diversity. It is feared some populations may not be able to recover from decline and that it may be necessary to reintroduce or supplement these plants.

Spreading Bellflower occurs on banks, roadsides and open outcrops and retains a strong association with forestry, rarely occurring far from ancient woodlands. Sites are generally sunny, with dry, well-drained and often low fertility soils.

In Britain, it mainly occurs in the southern Welsh Marches, with scattered sites elsewhere. It is a Western European rarity, being more common in meadows in Central and Eastern Europe into Russia, making British habitats look out of place.

Threats and Conservation

Spreading Bellflower's requirement for disturbance is linked to its seed germination; the decline in traditional woodland coppicing has resulted in lower germination rates and consequently there are fewer plants. Recent herbicide application along road banks and verges has added to the long-term decline in some populations.

Stag's-horn Clubmoss
Cnwp-fwsogl corn carw
Lycopodium clavatum

Welsh conservation status: Least concern

■ Before 1987
■ After 1987

An evergreen fern-relative which grows horizontally along the ground, with forked, upright shoots topped with yellowish-green cones. The leaves arise spirally along the stems and are bristle-like, tapering to a hair-like point which distinguish it from other British clubmosses.

The cones produce spores which, when ripe (June-September), are blown by the wind and distributed to colonise new sites. This may result in the surprise discoveries of new colonies, miles away from known sites, such as on rides in forestry plantations or on old spoil tips. Despite the spores, the majority of local spread is achieved vegetatively, by growth of the horizontal stem and the breaking up of its older sections.

This is a plant of grassy habitats, commonly in montane grasslands, as well as heathland and moors. It can tolerate very acid soil conditions.

Although once common throughout much of Britain, decline has resulted in the loss from most lowland sites so it is very sparse in Central and Eastern England. The European range is from Scandinavia south to Portugal, Italy and Bulgaria.

Threats and Conservation

Some lowland heathy sites were lost due to the conversion of sites to arable land, or colonisation by scrub. Other sites have suffered damage through excessive grazing, or the practice of heather burning. These pressures are continuing in some sites, but colonisation of new areas has also been observed. As such, populations are currently hard to follow and difficult to quantify, resulting in no conservation threats in Wales.

Reference: NHM (2011).

Thin-leaved Whitebeam
Cerddinen Fannau
Sorbus leptophylla

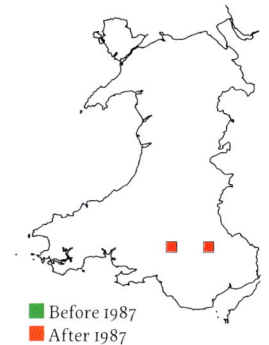

■ Before 1987
■ After 1987

Welsh conservation status: Endangered

A shrub, or, more rarely, a small tree up to 5m in height, with white inflorescences flowering in June. The fruits are dark red and characteristically longer than wide. The name 'Thin-leaved' does not seem to aptly describe this Whitebeam.

The growth habit of this plant seems to change according to its position and the subsequent levels of exposure or shading. Typically it grows against a rock face, with dense twiggy growth at its base and pendulous branches. However, in absence of such support, such as on cliff tops, trees with the same genetic make-up grow erect and with smaller leaves. In strongly shaded conditions, the trees do not flower.

Thin-leaved Whitebeam grows on cliff sides and tops, rooting into crevices and ledges. It prefers open, sheltered conditions, flowering and fruiting more successfully on open cliff-faces than those exposed on cliff-tops.

A rare Welsh endemic, this species is found only at Craig Rhiwarth and Craig y Cilau, with recent estimates of 29 and 45 trees respectively. Historical records elsewhere have been misidentified.

Threats and Conservation
There is no evidence of decline, recent population estimates actually being much increased in comparison to when populations were first found. This could be due to population increase, but could also be accounted for by increased survey quality, or awareness of the taxonomic differences from other Sorbus species. Its current threat status is due to the very low number of sites and individuals, making extinction due to chance events such as disease possible.

References: Rich et al. 2005. Rich et al. 2010.

Three-lobed Water-crowfoot
Crafanc-y-frân dridarn
Ranunculus tripartitus

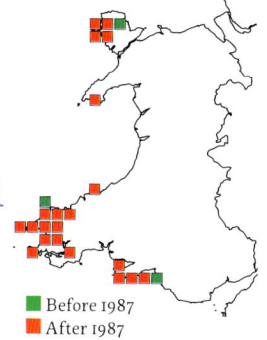

Before 1987
After 1987

Welsh conservation status: Vulnerable

A delicate, prostrate plant to 30cm long with leaves divided into three lobes (hence the name) and small, white petals up to 5mm long. Flowering occurs March–May.

In the 1990s, Three-lobed Water-crowfoot was thought to be very rare, having declined massively following the general post-war decline in grazing of heathlands. Plantlife then conducted targeted surveys, early in the year and in exactly the right habitat and found it was still present in many sites, albeit in very small quantities. Following the discovery that it had dormant seed banks hidden in the mud of seasonal pools, conservation work to open up the edges has seen a resurgence of 17 to 41 sites in two decades.

Three-lobed Water-crowfoot grows in and on the edges of shallow, seasonal pools and flowers early, before they dry up in the summer.

The pools are typically in grazed heathland, forming in ruts of trackways, on lane sides, in ditches and on seasonal pond margins.

It is mainly scattered near the coasts of Southern and Western Britain on lowland heaths, but decline has resulted in strong populations only remaining in South-west Wales, the New Forest and Lizard Peninsula. It occurs locally in Western Europe.

Threats and Conservation
The initial decline was caused by conversion of heathland to arable land, development of scrub and infilling of seasonal pools. Plantlife's work to reinstate grazing and restore pools on heathlands has been a real success, and, if such management is continued, this species should have a bright future.

Reference: Plantlife 2012. Rich 2004.

Toadflax-leaved St John's-wort Eurinllys Culddeilog

Hypericum linariifolium

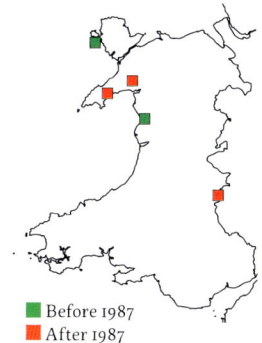

■ Before 1987
■ After 1987

Welsh conservation status: Vulnerable

A plant with greyish-green branching stems up to 40cm and narrow leaves. Each stem terminates in a flowering head: the flowers are five-petalled, bright yellow with a red tinge and black glands. Flowering is June–July.

This plant is typically a short-lived perennial and is quick to die when conditions are not right. To counteract this, large quantities of pollen are produced to increase the pollination chances, thus allowing vast numbers of seeds to be set when conditions are right. As a slow-growing, drought-tolerant species, it usually grows in open vegetation where competition is low and bare ground is maintained by summer drought.

Suitable open conditions are found on shallow soils and rocky slopes such as coastal cliffs, banks and valley sides and in open glades and rocks in woodland.

Populations are restricted to Western Britain, including Pwllheli, Gwynedd where it was first discovered in 1888. The species is endemic to the western coast of Europe, reaching its northern limit in Wales.

Threats and Conservation
The main threat to this species is overgrowth by surrounding scrub, which has resulted in the loss of some outlying populations, such as in Anglesey and Gwynedd. Scrub removal at some sites has resulted in increased population numbers and should be carried out at all sites at risk and subsequently be maintained. Over half of sites are protected in Sites of Special Scientific Interest or other managed sites.

Reference: Kay & John 1995. Jones 2011.

Touch-me-not Balsam
Ffromlys
Impatiens noli-tangere

Welsh conservation status: Endangered

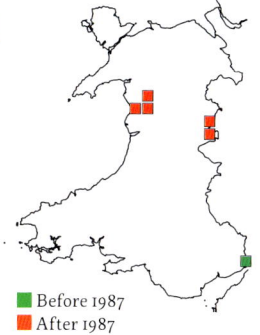

■ Before 1987
■ After 1987

A leafy annual plant to 60cm in height, flowering in July–September. The golden yellow flowers are spotted with brown and have a long spur at the base of the flower, which arcs downwards and then turns back through another 90°.

The native or introduced status of Touch-me-not Balsam varies across Britain, with records from places like South-east England usually being regarded as introductions. In North Wales and the Lake District, the presence of the Netted Carpet Moth in the same areas shows that the sites are, indeed, native. The Netted Carpet Moth is one of Britain's rarest moths, whose larvae feed exclusively on Touch-me-not Balsam, their rarity mirroring the rarity of the Balsam.

Touch-me-not Balsam usually grows in woodland areas with damp, nutrient-rich soils, often on streamsides and seepage areas in valleys. It flowers best in open or semi-shaded conditions.

Britain has small scattered populations north into Scotland. This includes two native Welsh sites at Dolgellau and on the Powys/Shropshire border, as well as the stronghold in the Lake District. British populations are considered westerly outliers of its main range, which is throughout much of Central Europe and into Asia.

Threats and Conservation

Following a historical decline, the range currently appears stable, but the populations are small, few and scattered. The Welsh border populations have decreased in recent decades following river management, which has increased flow rates making establishment harder and altering the moisture content of the soils. Potential conservation action would be worth considering.

Reference: Cumbria Wildlife 2011.

Tubular Water-dropwort
Cegiden bibellaidd
Oenanthe fistulosa

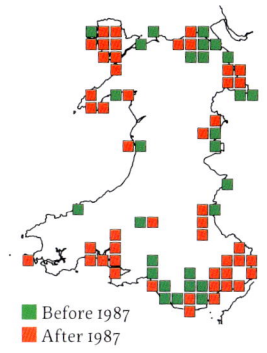

■ Before 1987
■ After 1987

Welsh conservation status: Vulnerable

A slender, waxy-bluish plant up to 80cm in height, with conspicuous, white, dome-shaped inflorescences, which also spreads by underground rhizomes. Flowering occurs July–September.

Plants from the genus *Oenanthe* are highly poisonous – the name *Oenanthe* means 'wine flower', due to the drunken-like effects produced when plants are eaten. Tubular Water-dropwort's bitter-taste and low toxin concentration means that no poisoning cases have been reported but its sweeter-tasting relative, Hemlock Water-dropwort, contains the same toxin in high concentrations and can be lethal.

Tubular Water-dropwort is a plant of wet grassland habitats such as flood plains, meadows and vegetation alongside waterways which are subject to flooding during winter.

The majority of British sites are in the east of England, with scattered local sites through to Scotland and Wales. This species also occupies much of Europe, the range reaching as far as South-west Asia and North-west Africa.

Threats and Conservation

Much decline has occurred, particularly in recent decades, resulting in loss of westerly sites in Wales, northerly sites in Scotland and throughout England. Much of this loss is a result of drainage of wet grasslands which are then re-seeded or converted to arable land. In order to prevent further decline, the remaining sites must be maintained and flood regimes allowed to continue.

Reference: Appendino et al. 2009.

Tufted Saxifrage
Tormaen siobynnog

Saxifraga cespitosa

Welsh conservation status: Critically Endangered

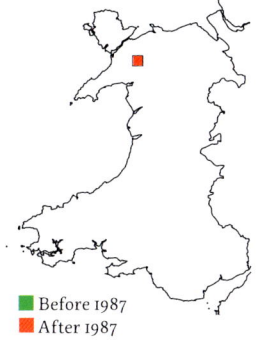

■ Before 1987
■ After 1987

A tiny plant with basal leaf rosettes that form tufts on rocky surfaces. The stems reach up to 10cm and have few or no leaves, but bear small, five-petalled, creamy flowers which open May–July.

Although the one site in Cwm Idwal National Nature Reserve in North Wales has been known since 1796, by 1975 numbers had dropped to only four plants, mainly due to collecting. Using seed from the remaining plants, a restocking programme was initiated, and, in 1978, plants and seeds were reintroduced near the original site. These have slowly declined since, and, once again, Tufted Saxifrage is facing an uncertain future. Oddly, it grows very well in cultivation.

Tufted Saxifrage is an arctic species, forming mats on mossy ledges, cracks and crevices typically above 600m altitude. Limited to this habitat, populations in Britain are present only on one Welsh and seven Scottish mountains.

Britain is the European southern limit of Tufted Saxifrage, the only other sites being in Norway. The arctic distribution continues into North America, including Alaska, Labrador and southwards through the Rockies.

Threats and Conservation
Conservation action is once again being directed at Tufted Saxifrage in Cwm Idwal. Studies are needed to see how much genetic variation remains in the populations in both the Welsh and Scottish sites (only one Scottish site has more than 30 plants) and whether this affects reproductive performance in the wild.

Reference: Parker 1981.

Upright Clover
Mellionen Unionsyth

Trifolium strictum

Welsh conservation status: Critically Endangered

■ Before 1987
■ After 1987

The green stem, up to 15cm in height, bears many sharply-toothed leaves and terminates in spherical flowering heads, the green sepals encasing pinky-purple and white petals. Flowering occurs May–June.

Upright Clover populations fluctuate from year to year, often in response to climatic conditions, as well as management. The seeds germinate at the end of summer and plants persist through the winter as leaf rosettes. In years with cold and wet winters, few plants survive the cold and populations are small. Population booms are seen in years following droughts, when open ground has been created.

It occurs on open, rocky, grazed grassland, often where soil layers are thin, such as on cliff slopes and rock outcrops.

Upright Clover is very rare in Britain and only occurs on the Lizard Peninsula in Cornwall and at Stanner Rocks National Nature Reserve, Powys. These are the northern-most sites of the global range, which extends southwards to Portugal to Turkey.

Threats and Conservation

Conservation management work at Stanner Rocks has recently boosted population numbers; soil disturbance and cutting dense grass with sheep shears has opened up the vegetation, resulting in a resurgence of the populations from buried seed. Populations at the Lizard Peninsula have also established in gorse scrub which was opened up by burning. Providing sites are managed, the British populations should survive.

Reference: Jones 2011.

Upright Spurge
Llaethlys Mynwy
Euphorbia stricta

Welsh conservation status: Near Threatened

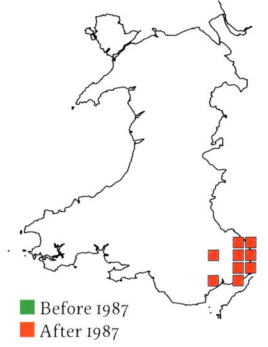

■ Before 1987
■ After 1987

A leafy plant up to 80cm in height. Many short, green stalks arise from the erect central stem to produce green-flowered inflorescences. Flowering is June–July.

Upright Spurge's seeds are produced in light-weight capsules, which are dropped when ripe. In addition to being dispersed via wind and water or remaining by the parent, ants may carry the capsules to their nests, where some seeds get discarded. The seeds are long-lived and their dormancy can be broken by disturbance, such as from increased light from coppicing woodlands. This can result in resurrection of populations in sites where it has not been seen for up to 40 years.

Its typical habitat is within limestone woodlands, including along tracks and hedgebanks. This provides the shelter required for successful growth, whilst allowing adequate light and open conditions for establishment.

Upright Spurge was discovered in 1773 in Monmouthshire. Its British distribution is still centred on the Wye Valley in Britain and the two largest populations are in Gloucestershire. Populations elsewhere (e.g. Cardiff) tend to be short-lived. It occurs widely in Western and Southern Europe, eastwards to Turkey.

Threats and Conservation
Sites for Upright Spurge have declined by nearly half and most remaining populations have fewer than 20 plants. This has mirrored changes in traditional woodland in recent decades, some habitat being lost to conversion to conifer plantations and some being altered by overgrowth of vigorous plants as traditional management decreases. The long-lived seed bank means conservation should be feasible if required conditions are reinstated, for example, through coppicing cycles.

Viper's-grass
Llys y wiber

Scorzonera humilis

Welsh conservation status: Vulnerable

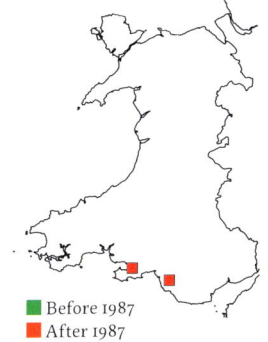

■ Before 1987
■ After 1987

The flowering buds of Viper's-grass are on stalks and, when arising from the rosette of long, thin leaves, look like the head of a snake rising up from the ground, hence the name. The yellow flowers, like those of a dandelion, open in May–June; it also has similar seeds, with parachutes.

Considering that Viper's-grass is common and widespread in Europe, it is surprisingly rare in Britain. Two sites in Dorset have been known for a long time (only one survives) and another in Warwickshire persisted until 1967. A real surprise was the discovery of two sites in Bridgend, South Wales in 1996 and 1997, both in species-rich, damp grasslands where they have obviously been for a long time.

The damp and marshy habitats in which it occurs are unimproved grasslands, meadows, heaths and fens. The species occurs throughout continental Europe.

Threats and Conservation

Viper's-grass is rare with only three remaining British sites, though at Bridgend and in Dorset the plants still number in their thousands. Overgrowth of other vigorous species must be prevented and this is achieved in Bridgend by grazing and weed-topping. In Dorset, the field is very over-grown and some grazing may be needed to prevent future decline.

Welsh Groundsel
Creulys Cymreig
Senecio cambrensis

Welsh conservation status: Critically Endangered

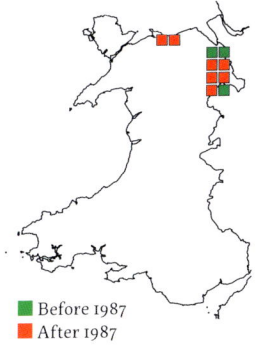

■ Before 1987
■ After 1987

A small annual herb, reaching around 50cm height. Its yellow flowers appear from May to October and its seeds (3-3.7mm long) are wind-dispersed.

Welsh Groundsel is an unusual example of a species which has evolved in Wales, when an infertile hybrid between Groundsel and Oxford Ragwort underwent chromosome doubling and became fertile. It is the only example of a sexually reproducing species endemic to Wales.

Welsh Groundsel is typically found growing on open, disturbed ground such as on roadsides, in gardens and on rubble. It does not compete well with other plants.

Surveys in 2010 show that it is now only found in 12 sites, all in North Wales, compared to the 27 sites recorded in the 1980s. A population found in Edinburgh in the 1980s has since disappeared and the decline looks set to continue within Wales.

Threats and Conservation
The last 20 years has seen a slow decline of Welsh Groundsel, primarily through habitat loss due to development, as well as mowing and herbicide application on roadsides. A sympathetic approach to the management of these areas could enable a more successful future.

References: Abbot et al. 1983, 2007. Ashton & Abbot 1992. Boyett 2011. Dines 2008. Ingram & Noltie 1995. Marren 1999. Rosser 1955.

Welsh Northern Marsh-Orchid
Tegeirian-y-gors Cymreig

Dactylorhiza purpurella var. *cambrensis*

Welsh conservation status: Vulnerable

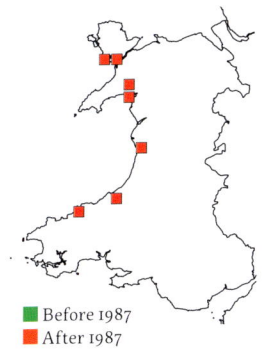

■ Before 1987
■ After 1987

A stunning plant with a dense red-purple inflorescence supported by a leafy stem which may reach up to 25cm. Flowering occurs June–July.

Debate and disagreement has often surrounded the identity of the variable marsh orchids, resulting in many names for this variety, and, consequently, a lack of certainty regarding historical records. Only two populations were known when it was originally described in 1961, but now at least seven sites are known, including Newborough Warren and Ynys-las.

True to its name, Welsh Northern Marsh-Orchid resides in wet habitats such as dune-slacks, fens, marshes, bogs and flushes.

Although the Northern Marsh-orchid is quite widespread in Northern and Western Britain, this Welsh variety is endemic to a handful of sites in West Wales. It can be locally frequent and number in the thousands.

Threats and Conservation

There is no evidence of decline in the Welsh Northern Mash-orchid, though the Northern Mash-orchid has declined elsewhere as a result of habitat destruction and drainage. The Welsh Northern Mash-orchid populations have more than doubled at Ynys-las since its first survey in 1958. The increase might be genuine or a result of increased knowledge about how to identify it. Marsh-orchid populations frequently contain an interesting mix of different plants, and, as such, conservation strategies may be best aimed at Marsh-orchid communities as a whole.

Reference: Chater et al. 2010.

Welsh Wood Stitchwort
Serenllys y Coed

Stellaria nemorum subsp. *montana*

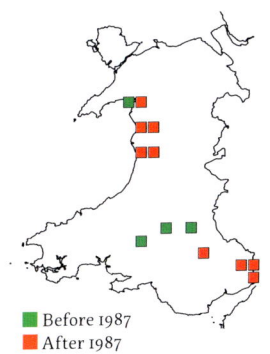

■ Before 1987
■ After 1987

Welsh conservation status: Critically Endangered

This delicate plant grows up to 60cm in height and has a hairy stem. The flowers have five white, deeply cleft petals. The flowering season is May–June.

Welsh Wood Stitchwort is a subspecies of the more widespread Common Wood Stitchwort. In Britain it is confined to Wales: it used to be recorded in Ceredigion, Gwynedd and Powys, but has only been seen recently in Monmouthshire and Carmarthenshire. The causes of this 70% decline are not understood, but could be related to lack of woodland management, and, in some sites at least, to hybridisation with Common Wood Stitchwort.

The main habitats are damp, shaded hedgerows, woodlands and streamsides.

In addition to its Welsh sites, this subspecies occurs across Europe to Spain, Sweden and Greece.

Threats and Conservation
More research is needed to understand why this plant is declining. Many woodlands have been neglected recently and have become overgrown, which might have contributed to the decline. There is also potential for further loss through hybridisation.

Reference: RA Jones 2010.

White Horehound
Llwyd y Cŵd

Marrubium vulgare

Welsh conservation status: Near Threatened

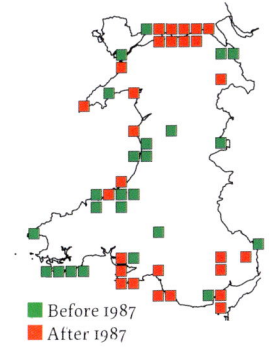

■ Before 1987
■ After 1987

The central leafy stem grows up to 60cm in height and the white flowers are in dense whorls at the leaf axils, giving the impression of intermittent 'balls' of flowers up the stem. Each flower has a two-lobed upper lip and three-lobed lower lip. Flowering occurs June–November.

White Horehound has long been used as a medicinal herb in Britain, where it was used to prepare tonics for sore throats, colds, coughs, asthma and bronchitis. More recent studies have shown that it can reduce cholesterol, triglycerides and plasma glucose levels, hence its use with *Cecropia obtusifolia* (the Trumpet tree) in medicines to treat Type 2 diabetes.

The historical medicinal use has cast doubt as to where it might be native in Britain. Many inland sites on disturbed ground and grasslands are considered to be escapes from cultivation. It is considered native on exposed cliff-tops and sloping grasslands, usually on the coast but sometimes inland on dry, mildly nutrient-rich soils; only the accepted native Welsh sites are plotted on the map.

White Horehound is scattered throughout Britain but is largely coastal in Wales. It occurs across much of Europe (except for the extreme north) and across into Asia and North Africa.

Threats and Conservation

The decline of White Horehound is difficult to quantify, as the introduced populations are relatively transient. Decline is certainly apparent within native populations and the coastal sites appear under threat from the decrease in cliff-top grazing, allowing increased growth of more competitive species. Potential action should involve the reinstatement of grazing, which has the added benefit of providing nutrients from animal dung.

References: Allen & Hatfield 2004. Herrera-Arellano et al. 2004.

Wild Asparagus
Merllys gorweddol

Asparagus prostratus

Welsh conservation status: Critically Endangered

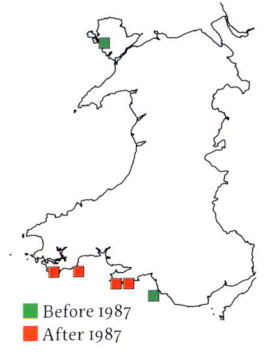

■ Before 1987
■ After 1987

The stems of this species grow prostrate along the ground to 70cm in length, with needle-like leaves. Yellow-cream flowers grow in the leaf axils and open May–June, but female plants are most conspicuous July-October when bearing red berries.

Only five sites are now known in Wales, mostly with fewer than 25 plants and little recruitment of young plants. Without conservation action these could be the last Wild Asparagus plants to exist in Wales: it already having been lost from Anglesey and three sites near Swansea. Transplants were sown to augment two populations in 2010 and 2011, and, so far, appear successful, with the new plants growing well.

Wild Asparagus is a coastal species, growing on well-drained sea-cliff grassland and slopes and on sand dunes, often in exposed sites.

In addition to the Welsh sites in Pembroke and Gower, Wild Asparagus occurs at one site in Dorset and in many sites in Cornwall around the Lizard. It is restricted to Western Europe, ranging from Germany to Northern Spain, including Eastern Ireland and the Channel Islands.

Threats and Conservation
Some decline has been attributed to visitor pressure, over-grazing or scrub encroachment, many sites suffering from trampling and erosion. The cliff sites on the Lizard are mostly quite inaccessible and stable. Recent searches have revealed a number of new sites.

References: Kay et al. 2001. Rich et al. 2000–2010.

Wild Cotoneaster
Cotoneaster y Gogarth

Cotoneaster cambricus

Welsh conservation status: Critically Endangered

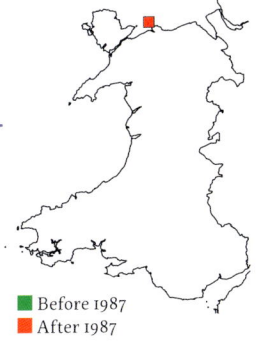

■ Before 1987
■ After 1987

A low-growing, spreading shrub whose stems may be up to 150cm long. The leaves are oval and greenish-blue. Flowering occurs April–June and the small, round, bright red fruits ripen in August.

After much debate, studies of DNA at the National Botanic Garden of Wales have finally shown that *C. cambricus* is a distinct species within the variable European *C. integerrimus* group. It has probably evolved since the last glaciation, after the Welsh populations became isolated from Europe. Sadly it has declined since it was first discovered in 1783 by John Wynne Griffith and now, with only six natural plants left, there is serious concern for how much longer this species will survive.

Wild Cotoneaster grows upon limestone rock ledges on Great Ormes Head at Llandudno in North Wales – the only place in the world where it occurs.

Threats and Conservation

Historically, in addition to the threats from over-grazing by sheep and feral goats, some decline was caused by local children who picked the plants and sold them for a few pence as tourist souvenirs. After studies showed it was not reproducing naturally, an intensive conservation programme is now in place, including exclusion of grazing animals, cultivation of plants and transplanting cuttings. There is hope for this special Welsh plant.

References: Fryer & Hylmo 1994. Kay & John 1995.

Wood Bitter-vetch
Ffacbysen chwerw
Vicia orobus

Welsh conservation status: Least concern

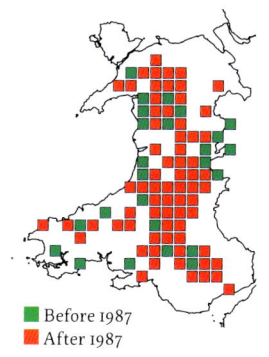

■ Before 1987
■ After 1987

The stems usually grow up to 60cm in height, supporting clusters of flowers like miniature sweet peas. The pretty flowers are about 15mm long, pinkish-white, flushed and veined with purple. Flowering occurs June–September.

The flowers are normally pollinated by bumblebees which reach inside the funnel-shaped flowers for nectar and get covered in pollen in the process. However, the small buff-tailed bumblebee, which is too small to reach the nectar from the front, makes small holes in the back of the flower and steals the nectar. These holes can then also be used by honeybees to steal more nectar. These nectar-robbers do nothing for pollination, or for the future of the plant!

Wood Bitter-vetch's habitat is rocky or wooded areas, such as banks, boulders and bushes, which are free from grazing to which it is intolerant, but with some prevention of scrub overgrowth. It also occurs occasionally in churchyards.

The British distribution is primarily in the west and north, the main strongholds being in Wales. The species is endemic to Western Europe, ranging from Spain and north to Norway.

Threats and Conservation
The primary concern for this species is the small size and isolation of many of the populations, combined with its poor ability to colonise new sites. Although this isolation poses no current issues, if genetic variation decreases, it may limit survival in the future. Potential conservation action could include linking or expansion of current populations, to allow for more genetic exchange.

Reference: Kay & John 1994.

Yellow Bird's-nest Cytwf

Hypopitys monotropa (=*Monotropa hypopitys*)

Welsh conservation status: Vulnerable

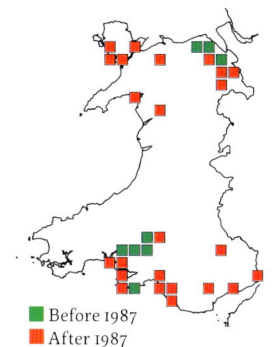

■ Before 1987
■ After 1987

A ghostly-looking species, pale brownish-white in colour with yellowed drooping flowers. It may reach up to 30cm in height and has many small, scale-like leaves. Flowering occurs June–July.

The colour of the plant is the result of it having no chlorophyll – the green pigment used by most plants to capture sunlight for photosynthesis. Rather than making its own energy, Yellow Bird's-nest is parasitic, using fungi associated with its roots to steal nutrients from neighbouring trees. The fungi are from the genus *Tricholoma* and are associated with Beech and Hazel trees in woodland or with Creeping Willow in dune slack habitats.

Yellow Bird's-nest typically grows in deeply shaded woodland and scrub on calcareous soils, its parasitic nature meaning it does not need light. Occasionally it will grow within acid woodlands beneath Pines. In Wales, its main habitat is in dune slacks.

Yellow Bird's-nest, first recorded in Britain in 1677, is scattered across much of Southern Britain, becoming rarer in the north and west. It occurs in much of the northern hemisphere, including Asia and North America.

Threats and Conservation
Research is required to see if the two different forms (which differ in the number of chromosomes and hairiness) are ecologically different or not. Little is known about its life cycle, the reasons for its decline, or why sometimes hundreds of plants appear for a few years and then disappear.

Yellow Centaury
Canrhi felen eiddil

Cicendia filiformis

Welsh conservation status: Vulnerable

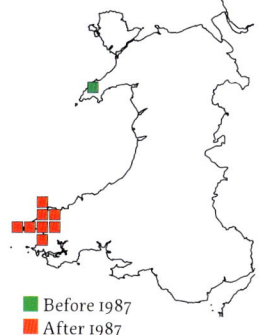

■ Before 1987
■ After 1987

At up to only 10cm high, this very slender short-lived plant has few leaves or flowers. The yellow petals spread open on sunny days from June to October, but it is probably mainly self-pollinated.

Yellow Centaury relies on disturbance, such as from winter flooding, trampling and grazing, to maintain open conditions and minimise competition from other plants. In good conditions, it grows in great quantities, only to disappear in subsequent seasons when these conditions are no longer present. Seeds can germinate at any time of year if it is damp enough.

Its habitats are typically damp and open and include heathland, pastures, pond edges, dune-slacks, cliffs, cart-tracks and woodland rides. In Holland it occurs on the edges of winter ice-skating ponds.

In Britain, Yellow Centaury mainly occurs in South-west England with scattered populations further north to Pembrokeshire (specially the St David's peninsula) and formerly Gwynedd. An overall decline has left the New Forest and Lizard Peninsula as strongholds, but a few new sites have been found recently in the west. It is distributed through Western Europe to North Africa and east to Turkey.

Threats and Conservation
Levelling of trackways and infilling of pools have been a major source of decline in the past, as have heathland destruction and cessation of grazing, which have also caused habitat loss. The majority of remaining populations now exist within protected sites. Management needs to ensure trackways remain lightly used and unimproved and that heathlands are kept open by grazing.

Yellow Whitlow-grass
Llysiau Melyn y Bystwyn
Draba aizoides

■ Before 1987
■ After 1987

Welsh conservation status: Near Threatened

Yellow Whitlow-grass is a perennial herb, whose leaves grow in cushion-forming rosettes at its base. Leafless stems grow 2-10cm in height and produce yellow, clustered flowers from late February–April.

Yellow Whitlow-grass is an alpine plant, which was probably initially widespread in lowland Europe after the last Ice Age, but became restricted to a few special sites as the climate warmed. One of these is in South Wales, far north of from the rest of its European distribution. This separation has genetically isolated the Welsh population, the plant being more uniform than its variable continental counterpart.

Yellow Whitlow-grass grows in crevices in limestone rocks and on walls. Being a plant intolerant of shading, these crevices prevent it being crowded out by neighbouring plants. The crevices are also typically south-facing, providing a higher exposure to the sun throughout the day.

Yellow Whitlow-grass is scattered across the Central and Southern European mountain ranges. In Wales it grows only along the Gower peninsula coastline and can be seen high on the ruined walls of Pennard Castle.

Threats and Conservation

Although not currently declining in Wales, this alpine is under threat from temperature rises associated with global warming. In addition, some accessible populations have previously been decimated by collectors, but most are now protected. The Pennard Castle population is at risk from repointing of the walls, but it was carefully looked after in 2013.

References: Kay & Harrison 1970. Rich 1991.

Six outstanding plant sites in Wales

The following six botanical sites are amongst the best in Wales and where some of the 101 rare plants, as well as many others, can be seen.

Brecon Beacons

Nestled amongst the mountain cliffs (some old red sandstone, some limestone), woodlands, meadows and waterfalls of the Beacons are Beacons Hawkweed, Bog Orchid, Globe Flower, Ley's Whitebeam, Least Whitebeam, Pillwort, Small-leaved Hawkweed and Thin-leaved Whitebeam.

Gower Peninsula

The extensive coast with limestones cliffs, sand dune systems, old commons and woodlands, coupled with the centre on more acidic soils and make this an incredibly rich area with Burnt Orchid, Dune Gentian, Juniper, Prickly Saltwort, Sea Stock, Wild Asparagus and Yellow Whitlowgrass.

Great Orme

The large limestone headland is packed with rare plants on the cliffs and grasslands, including Basil Thyme, Goldilocks Aster, Juniper, White Horehound and Wild Cotoneaster.

Snowdonia

Not surprisingly given its size, the Snowdonia mountain massif supports the largest number of rare Welsh plants including those of mountain cliffs, woodlands, grasslands, bogs and lakes such as Arctic Mouse-ear, Cumbrian Eyebright, Welsh Eyebright, Montane Eyebright, Ostenfeld's Eyebright, Floating Water-plantain, Glaucous Meadow-grass, Globe Flower, Holly Fern, Juniper, Killarney Fern, Marsh Clubmoss, Narrow-leaved Helleborine, Oblong Woodsia, Snowdon Lily, Snowdonia Hawkweed, Stag's-horn Clubmoss and Tufted Saxifrage.

Stanner Rocks

Stanner Rocks National Nature Reserve is a tiny outcrop of volcanic rock most of which has been quarried away, yet it supports an amazing concentration of rare plants including Early Star of Bethlehem, Perennial Knawel and Upright Clover.

Wye Valley

The meandering river, woodlands and rocky cliffs of the lower Wye Valley between Monmouth and Chepstow provides habitats for Lobed Maidenhair Spleenwort, Round-leaved Whitebeam and Upright Spurge.

Organisations involved in plant conservation in Wales

Botanical Society of Britain & Ireland

The Botanical Society of Britain and Ireland is an independent charitable membership organization for botanists interested in the wild flora of Britain and Ireland. It produces national Atlases and county Floras showing the distribution of plants, publishes the British and Irish Botany and newsletters and holds conferences and field meetings. It is the primary source of plant distribution data for Britain and Ireland.

Website: www.bsbi.org.uk

National Botanic Garden of Wales

The National Botanic Garden of Wales is dedicated to the research and conservation of biodiversity, to sustainability, lifelong learning and the enjoyment of the visitor. It is a charity which receives part funding from the Welsh Government. It was a founder member of the Rare Welsh Plants Project and grows many rare plants in the gardens.

Website: botanicgarden.wales

Amgueddfa Cymru

Amgueddfa Cymru provides a repository of information on the flora of Wales in the Welsh National Herbarium and the associated libraries and has undertaken research on conservation of the flora of Wales as a founder member of the Rare Welsh Plants Project.

Website: www.museum.wales

Natural Resources Wales

Natural Resources Wales is the Welsh Government's statutory advisor on sustaining natural beauty, wildlife and the opportunity for outdoor enjoyment in Wales and its inshore waters. In addition to the legislative responsibilities, it has funded research on plant conservation and was a founder member of the Rare Welsh Plants Project.

Website: www.naturalresourceswales.gov.uk

Plantlife

Plantlife is an independent charitable membership organization dedicated to dedicated to conserving wild plants and fungi in the places where they occur naturally. Plantlife carries out practical conservation work, manages nature reserves, influences policy and legislation, runs events and activities that connect people with their local wild plants and works with others to promote their conservation for the benefit of all.

Website: www.plantlife.org.uk/wales

References

Abbott, R.J., Noltie, H.J. & Ingram, R. (1983). The origin and distribution of S. cambrensis Rosser in Edinburgh. *Transactions of the Botanical Society of Edinburgh* 44: 103-106.

Abbott, R.J., Ireland, H.E. & Rogers, H.J. (2007). Population decline despite high genetic diversity in the new allopolyploid species *Senecio cambrensis* (Asteraceae). *Molecular Ecology* 16:1023-1033.

Aguraiuja, R. (2011). Reintroduction of the endangered fern species Woodsia ilvensis to Estonia: a long-term pilot study. *Biodiversity and Conservation* 20: 391-400.

Allen, D.A. & Hatfield, G. (2004). *Medicinal plants in folk tradition: An ethnobotany of Britain and Ireland.* Timber Press, Cambridge.

Appendino, G., Pollastro, F., Verotta, L., Ballero, M., Romano, A., Wyrembek, P., Szczuraszek, K., Mozrzymas, J.W. & Taglialatela-Scafati, O. (2009). Polyacetylenes from Sardinian *Oenanthe fistulosa*: A molecular clue to *risus sarsonicus*. *Journal of Natural Products* 72: 962-965.

Ashton, P.A, & Abbott, R.J. (1992). Multiple origins and genetic diversity in the newly arisen allopolyploid species, *Senecio cambrensis* Rosser (Compositae). *Heredity* 68: 25-32.

Bladwell, S., Dines, T. & Hobson, R. (2009 – revised). Action Plan for the Section 42, NERC Act, 2006 species. Accessed in 2011 at www.biodiversitywales.org.uk.

Boyett, L. (2011). The status of *Senecio cambrensis* Rosser, Welsh Groundsel. BSBI Recorder 15: 23.

Brysting, A.K. (2008). The arctic mouse-ear in Scotland – and why it is not arctic. *Plant Ecology and Diversity* 1: 321-327.

BSBI (2011). Botanical Society of Britain and Ireland, Species Accounts and Dossiers. Accessed at sppaccounts.bsbi.org.uk in 2011.

Campbell, V.V., Rowe, G., Beebee, T.J.C. & Hutchings, M.J. (2007). Genetic difference amongst fragrant orchids (*Gymnadenia conopsea* s.l.) in the British Isles. *Botanical Journal of the Linnean Society* 155: 349-360.

Chater, A.O., Allen, D.E., Preston, C.D. & Smith, P.A. (2010). *Flora of Cardiganshire.* Privately published. Aberystwyth.

Clapham, A.R., Tutin, T.G. & Moore, D.M. (1987). *Flora of the British Isles.* Third Edition. Cambridge University Press. Cambridge.

Clement, E.J. (2006). Could *Artemisia campestris* subsp. *maritima* be native? *BSBI News* 103: 4.

Cumbria Wildlife (2011). The Rare Netted Carpet Moth. *Cumbria Wildlife.* Accessed Dec 2011 at: www.cumbria-wildlife.org.uk/netted.html.

Davies, D. & Jones, A. (1995). *Enwau Cymraeg ar Blanhigion, Welsh Names of Plants.* National Museum of Wales. Cardiff.

Devos, N., Raspe, O., Jaquemart, A. & Tyteca, D. (2006) On the monophyly of *Dactylorhiza* Necker ex Nevski (Orchidaceae): is *Coeloglossum viride* (L.) Hartman a *Dactylorhiza*? *Botanical Journal of the Linnean Society* 152: 261-269.

Dines, T. (2008). *A vascular plant Red Data List for Wales*. Plantlife International. Salisbury.

Dines, T.D., Jones, R.A., Leach, S.J., McKean, D.R., Pearman, D.A., Preston, C.D., Rumsey, F.J. & Taylor, I. (2005). The Vascular Plant Red Data List for Great Britain. Species Status 7: 1-116. Joint Nature Conservation Committee, Peterborough.

Evans, S. (2005). Note on *Melittis melissophyllum* in Pembrokeshire, V.C. 45. BSBI Welsh Bulletin 76: 10-15.

Foley, M.J.Y (1990). The current distribution and abundance of *Orchis ustulata* L. in Southern England. *Watsonia* 18: 37-48.

Fryer, J. & Hylmo, B. (1994). The native British Cotoneaster – Great Orme Berry – renamed. *Watsonia* 20: 61-63.

Griffith, J.E. (1895). *The Flora of Anglesey and Caernarvonshire, with an account of their flowering plants, ferns and their allies, mosses, marine algae, lichens and hepaticae*. Nixon and Jarvis. Bangor

Godefroid, S., Van de Vyver, A., Stoffelen, P., Robbrecht, E. & Vanderborght, T. (2011). Testing the viability of seeds from old herbarium specimens for conservation purposes. Taxon 60: 565-569.

Herrera-Arellano, A., Aguilar-Santamaría, L., García-Hernández, B., Nicasio-Torres, P. & Tortoriello, J. (2004). Clinical trial of *Cecropia obtusifolia* and *Marrubium vulgare* leaf extracts on blood glucose and serum lipids in type 2 diabetics. Phytomedicine 11: 561-566.

Hutchinson, G. & Rich, T.C.G. (2005). Conservation of Britain's biodiversity: *Hieracium radyrense* (Asteraceae), Radyr Hawkweed. *Watsonia* 25: 403-407.

Ingram, R. & Noltie, H.J. (1995). Biological flora of the British Isles. *Senecio cambrensis* Rosser. *Journal of Ecology* 83: 537-546.

Ignacuik, R. & Lee, J.A. (1980). The germination of four annual strandline species. *New Phytologist* 84: 581-591.

Jaume, P., Garcia, S., Garnatje, T. & Vallès, J. (2009). Changes in genome size in a fragmented distribution area: the case of *Artemisia crithmifolia* L. (Asteraceae, Anthemideae). *Caryologia* 62: 52-160.

Jermy, A.C., Simpson, D.A., Foley, M.J.Y. & Porter, M.S. (2007). *Sedges of the British Isles*, B.S.B.I Handbook No. 1. 3rd edition. Botanical Society of the British Isles. London.

Jersáková, J., Malinová, T., Jeřábková, K. & Dötterl, S. (2011). Biological Flora of the British Isles: *Pseudorchis albida* (L.) Á. & D. Löve. *Journal of Ecology* 99: 1282–1298.

Jones, B. (2001). Seed plants. In, Rhind, P. & Evans, D. eds., *The plant life of Snowdonia*. pp. 1-36. Gomer Press. Llandysul.

Jones, B. (2010). Snowdon Lily (*Lloydia serotina*). Snowdonia National Park Authority (website). Accessed Dec 2011 at www.eryri-npa.gov.uk/conserving/biodiversity-in-snowdonia/priority-species/plants/snowdon-lily-lloydia-serotina.

Jones, B. & Gliddon, C. (1999). Reproductive biology and genetic structure in *Lloydia serotina*. Plant Ecology 141: 151-161.

Jones, B., Gliddon, C. & Good, J.E.G. (2001). The conservation of variation in geographically peripheral populations: *Lloydia serotina* (Liliaceae) in Britain. Biological Conservation 101: 147–156.

Jones, D. (2003). *Welsh Wildlife*. Y Lolfa Cyf. Talybont, Ceredigion.

Jones, R.A. (2010). Welsh Wood Stitchwort (*Stellaria nemorum* subsp. *montana*) under threat. *BSBI Welsh Bulletin* 86: 13-15.

Jones, R.A. (2011). Notes on rare plant fluctuation and management at Stanner Rocks NNR. *BSBI Welsh Bulletin* 87: 20-23.

Karin, A. & Karlson, B. (1990). Chemical and ethological studies of pollination in the genus *Ophrys* (Orchidaceae). *Phytochemistry* 29: 1359-1387.

Karlsson, L.M., Ericsson, J. & Milberg, P. (2006). Seed dormancy and germination in the summer annual *Galeopsis speciosa*. *Weed Research* 46: 353-361.

Kay, Q.O.N. & Harrison, J. (1970). Biological Flora of the British Isles, *Draba aizoides* L. *Journal of Ecology* 58: 877-888.

Kay, Q.O.N. & John, R. (1994). Population genetics and demographic ecology of some scarce and declining vascular plants of Welsh lowland grassland and related habitats. *Countryside Council for Wales Science Report, No. 93.* Countryside Council for Wales.

Kay, Q.O.N. & John, R.F. (1995). The conservation of scarce and declining plant species in lowland Wales: Population genetics, demographic ecology and recommendations for future conservation in 32 species of lowland grassland and related habitats. *Countryside Council for Wales Science Report, No. 110.* Countryside Council for Wales.

Kay, Q.O.N., Davies, E. W. & Rich, T.C.G. (2001). Taxonomy of the Western European endemic *Asparagus prostratus* (*A. officinalis* subsp. *prostratus*) (Asparagaceae). *Botanical Journal of the Linnean Society* 137: 127-137.

Khojasteh-Bakht, S.C., Chen, W., Koenigs, L.L., Peter, R.M. & Nelson, S.D. (1999). Metabolism of (R)-(+)-Pulegone and (R)-(+)-Menthofuran by Human Liver Cytochrome P-450s: Evidence for Formation of a Furan Epoxide. *Drug Metabolism and Disposition* 5: 574-580.

Maad, J. & Nilsson, L.A. (2004). On the mechanism of floral shifts in speciation: gained pollination efficiency from tongue- to eye-attachment of pollinia in *Platanthera* (Orchidaceae). *Botanical Journal of the Linnean Society* 83: 481-495.

Mahboubi, M. & Haghi, G. (2008). Antimicrobial activity and chemical composition of *Mentha pulegium* L. essential oil. *Journal of Ethnopharmacology* 119: 325-327.

Marren, P. (1999). *Britain's rare flowers.* T. & A.D. Poyser Ltd. London.

McCosh, D. & Rich, T. (2018). *Atlas of British and Irish Hawkweeds* (Pilosella L. and Hieracium L.). Botanical Society of Britain and Ireland, Harpenden.

Metherell, C. & Rumsey, F.J. (2018). *Eyebrights (Euphrasia) of the UK and Ireland.* BSBI Handbook no. 18. BSBI, Bristol.

Moore, S.J. (2009). Distribution and current status of three Welsh endemic Hawkweeds: *Hieracium breconicola, Hieracium britannicoides* and *Hieracium subbritannicum*. MSc. thesis, University of Glamorgan, Pontypridd, September 2009.

Moughan, J. & de Vere, N. (2012). Conservation of *Salvia pratensis*: Meadow Clary/Saets y Waun in Wales. Unpublished report from National Botanic Garden of Wales to Countryside Council for Wales.

Murphy, R.J. (2009). *Fumitories of Britain and Ireland.* BSBI Handbook no. 12. Botanical Society of the British Isles. London.

NHM (2011). Taxonomy; *Lycopodium clavatum* (wolf paw clubmoss). Within 'Nature Online' of the Natural History Museum, London. Accessed December 2011 at www.nhm.ac.uk/nature-online/species-of-the-day/scientific-advances/industry/lycopodium-clavatum/taxonomy/index.html.

Parker, D.M. (1981). The re-introduction of *Saxifraga cespitosa* to north Wales. pp. 506-508 in Synge, H. (ed.), The biological aspects of rare plant conservation. Proceedings of International Conference, King's College, Cambridge, 14-19 July 1980. Chichester. Wiley.

Pellmyr, O. (1989). The cost of mutualism: interactions between *Trollius europaeus* and its pollinating parasites. *Oecologia* 78: 53-59.

Preston, C.D., Pearman, D.A. & Dines, T.D. (2002). *New Atlas of the British and Irish Flora, An atlas of the vascular plants of Britain, Ireland, the Isle of Man and the Channel Islands*. Oxford University Press. Oxford.

Plantlife (2011). Species Accounts, Briefing Pages and Dossiers. Accessed on October 2011 at www.plantlife.org.uk/wild_plants/plant_species/

Plantlife (2012). Wild Plant Horizons: Taking forward the Global Strategy for Plant Conservation 2011-2020. *Report on Behalf of Plant Link*. Accessed April 2011 at www.plantlife.org.uk/uploads/documents/GSPC_Template_v4.pdf

Pryce, K. & Pryce, R. (2011). Monitoring *Melittis melissophyllum* (Bastard Balm) in Carmarthenshire. *BSBI Welsh Bulletin* 87: 23-26.

Rebane, M. (2010). UKBAP Priority Habitat Action Plan – Upland Flushes, Fens and Swamps. *Natural England*. Published online www.biodiversitywales.org.uk/content/uploads/documents/SG Meetings/TaskFinish/Upland/UKBAP Priority HAP Upland Flushes Fens and Swam_2.doc. Accessed October 2011.

Rich, T.C.G. (1991). *Crucifers of Great Britain and Ireland. B.S.B.I Handbook No.6*. Botanical Society of the British Isles. London.

Rich, T.C.G. (1997). The management of semi-natural lowland grassland for selected rare and scarce vascular plants: a review. *English Nature Research Reports*. English Nature.

Rich, T.C.G. (2003). Conservation of Britain's biodiversity: *Hieracium snowdoniense* (Asteraceae), Snowdonia hawkweed. *Watsonia* 24: 513-518.

Rich, T.C.G. (2004). Vascular Plants. In, Department of Biodiversity and Systematic Biology 2004. *Biodiversity Wales, Species of conservation or special interest to Wales*. National Museums and Galleries of Wales. Cardiff.

Rich, T.C.G. (2005). Could *Centaurium scilloides* (L. f.) Samp. (Gentianaceae), Perennial Centaury, have colonised Britain by sea? *Watsonia* 25: 397-401.

Rich, T.C.G. & Karran, A.B. (2006). Floristic changes in the British Isles: comparison of techniques for assessing changes in frequency of plants with time. *Botanical Journal of the Linnean Society* 152: 279-301.

Rich, T.C.G. & McVeigh, A. (2019 in press). *Gentians of Britain and Ireland*. BSBI Handbook 19. Botanical Society of Britain and Ireland, Harpenden.

Rich, T.C.G. & Pryor, K.V. (2003). *Galeopsis segetum* Neck. (Lamiaceae), Downy Hemp-nettle: native or introduced in Britain? *Watsonia* 24: 401-411.

Rich, T. C. G., Bennallick, I.J., Cordrey, L., Kay, Q.O.N., Lockton, A. & Rich, L.K. (2002). Distribution and population sizes of Asparagus prostratus Dumort., Wild Asparagus, in Britain. *Watsonia* 24: 183-192.

Rich, T.C.G., Cordrey, L., Jones, A. & Leach, S. (2010). Wild about asparagus. *British Wildlife* 21: 305-311.

Rich, T.C.G., Donovan, P., Harmes, P., Knapp, A., McFarlane, M., Marrable, C., Muggeridge, N., Nicholson, R., Reader, M., Reader, P., Rich, E. & White, P. (1996). *Flora of Ashdown Forest.* Sussex Botanical Recording Society. Sussex.

Rich, T.C.G., Houston, L., Robertson, A. & Proctor, M. (2010). *Whitebeams, Rowans and Service Trees of Britain and Ireland. A monograph of British and Irish Sorbus L.* BSBI Handbook no. 14. Botanical Society of the British Isles. London.

Rich, T.C.G., Jones, R.A. & Lockton, A. (2000). On the Anglesey records for *Asparagus prostratus* Dumort., Wild Asparagus. *BSBI Welsh Bulletin* 67: 11-12.

Rich, T.C.G., Lambrick, C.R. & McNab, C. (1999). Conservation of Britain's biodiversity: *Salvia pratensis* L. (Lamiaceae), Meadow Clary, 1994-1996. *Watsonia* 22: 405-411.

Rich, T.C.G., Motley, G.S. & Kay, Q.O.N. (2005). Population sizes of three rare Welsh endemic *Sorbus* species (Rosaceae). *Watsonia.* 25: 381-388.

Rich, T.C.G., Rich, L.K., Evans, S.B., Evans, A.E. & Hopkins, F. (2006). Vegetation and habitats of the western European endemic *Asparagus prostratus* Dumort. (Asparagaceae), Wild Asparagus. Pp. 231-241, in Leach, S. J., Page, C. N., Peytoureau, Y. & Sanford, M. N., *Botanical Links in the Atlantic Arc.* BSBI Conference Report no. 24. BSBI/English Nature, London.

Rich, T.C.G., McVeigh, A. & Stace, C.A. (2018). New taxa and new combinations in the British flora. *Edinburgh Journal of Botany* 1-8. doi: 10.1017/S0960428618000288.

Rich, T.C.G., Houston, L. & Tillotson, A.C. (2019). *Sorbus* diversity in the Wye Valley Woodlands SAC, Wales. NRW Evidence Report Series. Report No: 332. Natural Resources Wales, Cardiff.

Rix, E.M. & Woods, R.G. (1981). *Gagea bohemica* (Zauschner) J.A. & J.H. Schultes in the British Isles and a general review of the *G. bohemica* species complex. *Watsonia* 13: 265-270.

Rosser, E.M. (1955). A new British species of *Senecio. Watsonia* 3: 228-232.

Rumsey, A. (2010). *Cephalanthera longifolia* (l..) Fritsch. *Plantlife International.* Published online, accessed October 2011: www.plantlife.org.uk/uploads/documents/Cephalanthera_longifolia_species_dossier.pdf.

Salisbury, E.J. (1961). *Weeds & Aliens.* New Naturalist Series, Collins, London.

Sawtschuk, J. (2006). Conservation of endemic *Hieracium* species in the British Isles and assessment of four Welsh species: *Hieracium pachyphylloides, Hieracium pseudoleyi, Hieracium rectulum* and *Hieracium cambricogothicum.* ESEB Masters Thesis, Université de Rouen, France.

Sell, P. & Murrell, G. (2006). *Flora of Great Britain and Ireland, volume 4. Campanulaceae – Asteraceae.* Cambridge University Press. Cambridge.

Shewring, M. & Rich, T.C.G. (2010). Conservation of Britain's biodiversity: status of the Welsh endemic *Hieracium subminutidens*, Llanwrytyd Hawkweed (Asteraceae). *Watsonia* 28:145-149.

Silverside, A.J. (1991). The identity of *Euphrasia officinalis* L. and its nomenclatural implications. *Watsonia* 18:343-450.

Smith, P.H. (2008). Increase in *Artemisia campestris* subsp. *maritima* at Crosby sand dunes, Merseyside. *BSBI News* 107: 28-29.

Smith, P.H. & Lockwood, P.A. (2011). Grazing is the key to the conservation of *Gentianella campestris* (L.) Börner (Gentianaceae): evidence from the north Merseyside sand-dunes. *New Journal of Botany* 1: 127-136.

Smith, P.H. & Wilcox, M.P. (2006). *Artemisia campestris* subsp. *maritima*, new to Britain, on the Sefton Coast, Merseyside. *BSBI News* 103: 3.

Stace, C.A. (2019). *New Flora of the British Isles*. 4th edition. C & M Floristics, Middlewood Green.

Stewart, A., Pearman, D.A. & Preston, C.D. (1994). *Scarce Plants in Britain*. JNCC. Peterborough.

Twibell, J.D. (2007). On the status of *Artemisia campestris* ssp. *maritima* as a native. *BSBI News* 104: 21-23.

Wigginton, M.J. (1999). *British Red Data Books, 1 vascular plants*. 3rd ed. Joint Nature Conservation Committee. Peterborough.

Willems, J.H. & Melser, C. (1998). Population dynamics and life-history of *Coeloglossum viride* (L.) Hartm.: an endangered orchid species in The Netherlands. *Botanical Journal of the Linnean Society* 126: 83-93.

Acknowledgements

This book arose from the Welsh Rare Plants Project 1998-2013, which aimed to help conserve threatened Welsh plants by providing a scientific basis for their conservation. Here we make the research more generally accessible and tell the stories of the plants and what we are doing to conserve them supported by some stunning photographs. The text was originally drafted in 2011 by Lauren during her professional training year as part of her degree at Cardiff University but has been long-delayed in publication so we have updated it with some recent work and names following the latest standard British flora (Stace 2019).

We would like to thank Trevor Dines of Plantlife, Andy Jones of Natural Resources Wales and Natasha de Vere of the National Botanic Garden of Wales for their help and support, and Helen Cleal for her editorial assistance.

We would like to thank the following photographers for permission to use their pictures and help with data: Ian Bonner, Sam Bosanquet, Michael Chalk, Arthur Chater, John Crellin, Liz Dean, Zoe Devlin, Trevor Dines, Trevor Evans, Peter Foulkes, Tom Fowler, Peter Garner, Chris Gibson, Fiona Gomersall, Charles Hipkin, Richard Lansdown, Peter Llewellyn, Brendan Marrinan, John O'Reilly, George Peterken, Sharon Pilkington, Kath Pryce, Richard Pryce, John Richards, Lliam Rooney, Fred Rumsey, Andy Shaw, Cath Shelswell, Barry Stewart, Kate Thorne, Geoff Toone, Mike Waller, Sarah Whild, Delyth Williams, Julian Woodman, Ray Woods and Jean Wynne-Jones. We would also like to thank photographers who offered pictures which we were unable to use.

Photo credits

We would like to thank the following photographers for use of their images. The numbers refers to the species number as listed in the contents.

Sam Bosenquet: 2, 8, 52, 56, 90, 98 (distant), 99.

Michael Chalk: 6.

John Crellin: 3, 11, 21, 33, 34, 35, 53, 57, 58, 61, 62, 72, 78, 81 (close-up), 86 (close-up), 95, 98 (close-up), 101.

Zoe Devlin: 55.

Trevor Dines: 14, 66, 84.

Peter Foulkes: 82.

Peter Garner: 74, 87.

Chris Gibson: 8, 69, 71.

Fiona Gomersall: 45.

Charles Hipkin: 12, 16, 28, 41, 70.

Richard Lansdown: 13, 18, 86 (distant).

Peter Llewellyn: 39, 73, 80, 88.

Brendon Marrinan: 51.

John O'Reilly: 54.

George Peterken: 91.

John Pilkington: 10.

Richard Pryce: 23, 28 (close-up), 75.

Tim Rich: 1, 5, 7, 9, 15, 19, 20, 29, 30, 32, 36, 43, 47, 48, 49, 50, 51, 54, 59, 63, 64, 65, 68, 76, 79, 83, 96, 97, 100.

John Richards: 42.

Liam Rooney: 22.

Fred Rumsey: 55, 57.

Andy Shaw: 4, 31, 38, 60, 67, 85, 89.

Mike Waller: 77, 93.

Sarah Whild: 40, 46, 92.

Julian Woodman: 94.

Ray Woods: 44, 81 (distant).

Goronwy Wynne: 37.

Jean Wynne-Jones: 45 (close-up).

Lauren Marrinan

Tim Rich

Lauren studied Ecology at Cardiff University alongside working in scientific engagement at Amgueddfa Cymru. She now works as a Biology teacher in Gloucestershire, where she continues to enjoy walking in the countryside and appreciating British flora and fauna.

Dr Tim Rich trained as an ecologist at Lancaster University and specialised in botany. He ran the Welsh National Herbarium for 17 years and is a national expert with particular interest in whitebeams and hawkweeds. He is currently working in applied ecology.

Index

A

Aegonychon purpureocaeruleum	127
Artemisia campestris subsp. *maritima*	39
Asparagus prostratus	193
Asplenium trichomanes subsp. *pachyrachis*	103
Aster linosyris	79
Aurfanadl Blewog	83

B

Blodyn-ymenyn yr ŷd	27
Blysmus compressus	63
Brenhinllys y maes	13
Brymlys	117
Bupleurum tenuissimum	149

C

Caldrist gulddail	111
Campanula patula	163
Camri	25
Canrhi barhaol	119
Canrhi felen eiddil	201
Carex divisa	33
Carex muricata subsp. *muricata*	93
Cegiden bibellaidd	175
Centaurea cyanus	29
Centaurium scilloides	119
Centaurium portense	119
Cephalanthera longifolia	111
Cerastium nigrescens	11
Cerddinen Fannau	167
Cerddinen Mynwy	137
Cerddinen Leiaf	95
Cerddin Ley	99
Chamaemelum nobile	25
Chweinllys Arfor	161
Cicendia filiformis	201
Clinopodium acinos	13
Clust-y-llygoden ogleddol	11
Clychlys ymledol	163
Cnwp-fwsogl corn carw	165
Cnwp-fwsogl y gors	105
Coeloglossum viride	73
Coreffros Cymreig	57
Corfrwynen	41
Cor-redynen hirgul	113
Corsfrwynen arw	63
Cotoneaster cambricus	195
Cotoneaster y Gogarth	195
Crafanc-y-frân dridarn	169
Creulys Cymreig	185
Crib Gwener	143
Cronnell	77
Crwynllys Cymreig	37
Crwynllys cynnar	47
Crwynllys y maes	61
Cytwf	199

D

Dactylorhiza purpurella var. *cambrensis*	187
Dianthus armeria	31
Dinodd Parhaol	121
Dinodd unflwydd	9
Draba aizoides	203
Duegredynen gwallt y forwyn	103
Dŵr-lyriad nofiadwy	65
Dyfrllys camleswellt	81

E

Effros Chwareog Gwalltog	53
Effros Ostenfeld	55
Effros y calch	51
Effros yr Wyddfa	52
Eleocharis parvula	43
Eriophorum gracile	147
Euphorbia stricta	181
Euphrasia anglica	53
Euphrasia cambrica	57
Euphrasia officinalis subsp. anglica	53
Euphrasia ostenfeldii	55
Euphrasia pseudokerneri	51
Euphrasia rivularis	52
Euphrasia officinalis subsp. monticola	54
Eurinllys Culddeilog	171

F

Ffachysen chwerw	197
Ffromlys	173
Fioled welw	115
Fumaria purpurea	129

G

Gagea bohemica	49
Gagea serotina	157
Galatella linosyris	79
Galeopsis angustifolia	133
Galeopsis segetum	35
Galeopsis speciosa	91
Gefell-lys y fignen	59
Genista pilosa	83
Gentianella amarella subsp. anglica	47
Gentianella amarella subsp. occidentalis	37
Gentianella anglica	47
Gentianella campestris	61
Gentianella uliginosa	37
Glas yr ŷd	29
Gludlys amryliw	151
Gold y Môr	79
Gwenynog	15
Gweunwellt llwydlas	75
Gweunwellt Oddfog	21
Gymnadenia borealis	71
Gymnadenia conopsea	69
Gymnadenia densiflora	70

H

Haidd y morfa	139
Hammarbya paludosa	19
Heboglys Dyfed	45
Heboglys Llanwrtyd	101
Heboglys Mân-ddail	153
Heboglys Radyr	131
Heboglys y Bannau	17
Heboglys Yryri	159
Helys pigog	125
Hesgen Bigog Gynnar	93
Hesgen ranedig	33
Hieracium angustatiforme	153
Hieracium breconicola	17
Hieracium radyrense	131
Hieracium rectulum	45
Hieracium snowdoniense	159
Hieracium subminutidens	101
Hordeum marinum	139

Hypericum linariifolium	171	**N**	
Hypopitys monotropa	199	Neotinea ustulata	23

I

Impatiens noli-tangere	173	**O**	
		Oenanthe fistulosa	175
		Ophrys insectifera	67

J

Juncus capitatus	41	**P**	
Juniperus communis	87	Paladr trwyddo eiddilddail	149

L

		Pelenllys	123
Lili'r Wyddfa	157	Penigan y porfeydd	31
Liparis loeselii	59	Pilularia globulifera	123
Lithospermum		Platanthera bifolia	97
purpureocaeruleum	127	Plu'r Gweunydd Eiddil	147
Llaethlys Mynwy	181	Poa bulbosa	21
Llwyd y Cŵd	191	Poa glauca	75
Llys y wiber	183	Polystichum lonchitis	85
Llysiau Melyn y Bystwyn	203	Potamogeton compressus	81
Luronium natans	65	Potentilla rupestris	135
Lycopodiella inundata	105	Pseudorchis albida	155
Lycopodium clavatum	165	Pumnalen y graig	135

M

R

Maenhad Meddygol	127	Ranunculus arvensis	27
Marrubium vulgare	191	Ranunculus tripartitus	169
Matthiola sinuata	141	Rhedynen gelyn	85
Melittis melissophyllum	15	Rhedynen wrychog	89
Mellionen Unionsyth	179	Rumex rupestris	145
Mentha pulegium	117		
Merllys gorweddol	193	**S**	
Merywen	87	Saets y Waun	109
Monotropa hypopitys	199	Salsola kali subsp. kali	125
Murwyll arfor	141	Salvia pratensis	109
Mwg y ddaear glasgoch	129	Saxifraga cespitosa	177
		Sbigfrwynen Morafon	43
		Scandix pecten-veneris	143

Scleranthus annuus	9
Scleranthus perennis subsp. perennis	121
Scorzonera humilis	183
Senecio cambrensis	185
Seren y Creigiau	49
Serenllys llwydlas	107
Serenllys y Coed	189
Silene gallica	151
Sorbus eminens	137
Sorbus leptophylla	167
Sorbus leyana	99
Sorbus minima	95
Stellaria nemorum subsp. montana	189
Stellaria palustris	107

T

Tafolen y traeth	145
Tegeirian bach gwyn	155
Tegeirian bach y gors	19
Tegeirian llosg	23
Tegeirian llydanwyrdd bach	97
Tegeirian pêr	69
Tegeirian pêr gogleddol	71
Tegeirian pêr y gors	70
Tegeirian y broga	73
Tegeirian y clêr	67
Tegeirian-y-gors Cymreig	187
Tephroseris integrifolia subsp. maritima	161
Torfagl Mynyddog	54
Tormaen siobynnog	177
Trichomanes speciosum	89
Trifolium strictum	179
Trollius europaeus	77

V

Vicia orobus	197
Viola lactea	115

W

Woodsia ilvensis	113

Y

Y Benboeth	35
Y Benboeth amryliw	91
Y Benboeth gulddail	133
Y feidiog ddi-sawr	39

What you can do to help

Please join Plantlife or the Botanical Society of Britain & Ireland and help contribute to their work; please contact them via their websites (see pages 208-209).

If you see any of our rare Welsh plants, DO NOT pick them! Please submit the records to Wales' Biodiversity Information & Reporting Database Aderyn (aderyn.lercwales.org.uk) with a photograph and details of where you found it and how many plants.